W9-BNN-649

**"WE ARE . . . AT THE BRINK OF A NEW AGE:
WHAT SOME EXPERTS CALL 'CATASTROPHIC TERRORISM.' "**

This book will tell you what that will mean when the inevitable happens, and what we as a nation can and must do now to protect ourselves, our families, and the basic elements of our society that bioterrorism puts at risk.

You will also learn why the billions being spent to combat terrorism miss the mark when it comes to fighting microbial terror. It will show why biological terrorism is becoming so much more likely, and examines the challenges of government response to it.

The book you hold in your hands is a manifesto, urging each of us to prepare for the inevitable or pay a stupendous toll in human lives. It lays out a prescription for survival—for our nation, and for us all.

Delta
Trade Paperbacks

WHAT AMERICA NEEDS TO KNOW
TO SURVIVE THE COMING
BIOTERRORIST CATASTROPHE

LIVING TERRORS

Michael T. Osterholm, Ph.D., M.P.H. & John Schwartz

A Delta Book
Published by
Dell Publishing
A division of Random House, Inc.
1540 Broadway
New York, New York 10036

Cover photograph by Harald Sund/Image Bank

Cover design by Craig DeCamps

BOOK DESIGN BY GLEN M. EDELSTEIN

ISBN 0-385-33481-8

Reprinted by arrangement with Delacorte Press, a division of
Random House, Inc.

Manufactured in the United States of America

Published simultaneously in Canada

October 2001

10 9 8 7 6 5 4 3

BVG

To Donald A. Henderson, who, more than twenty years ago, led mankind's greatest public health and medical accomplishment—the eradication of smallpox—and who has courageously entered the fight again to prevent its horrible return

CONTENTS

ACKNOWLEDGMENTS

This book is the work of many minds and helpful hands. We both owe a debt of immeasurable gratitude to our editor, Tom Spain, whose shared vision, tenacious support, and critical insight and suggestions were critical to the birth of this book. We also owe our sincere thanks to Andrea Nicolay, Tom's assistant, for her ongoing support.
—Michael and John

I have spent most of my career reassuring the public, policy makers, and colleagues that bacterial meningitis is not easy to transmit, HIV won't mysteriously spread through the air, and that immunizations are generally quite safe and very effective. While that's still true, many of our recent battles with infectious agents have become more complex and challenging. Nothing has prepared us, however, for the potential hell of a biological terrorism attack. For this reason, I felt compelled to write this book. I owe many who have helped me along the way.

I am profoundly grateful to John Schwartz, my co-author. I've known John for many years as one of the best science reporters in the business. The opportunity to collaborate with him on this project was a gift. His clear and concise thinking, as well as his tough standards for confirming the facts, were enormously important to the validity of what we've printed. His friendship has been priceless.

I thank Jonathon Lazear, my agent, for his encouragement and for tirelessly championing this effort. Without him, this book would not exist. Christi Cardenas, also of the Lazear Agency, was a constant voice of support.

Mark and Carolyn Olshaker and Ann Hennigan make up a big part of the Mindhunters. My deepest and most sincere gratitude go to them for working with me on several articles on this topic. They are the dearest of friends. Most of all, I owe Mark much for the countless hours of interviews and incredible work product that is reflected in this effort.

There is no group of individuals who deserve more credit for responding to the threat of bioterrorism than those at the Johns Hopkins University Center for Civilian Biodefense Studies. This includes D. A. Henderson, John Bartlett, Thomas Inglesby, Tara O'Toole, and Monica Schoch-Spana. Thank you for your ever-present support and guidance.

I want to thank my fellow state and city epidemiologists who have struggled with this issue, and in turn, have taught me a lot. This includes Marci Layton, Annie Fine, and Dennis Perrotta.

There are a number of professional colleagues who have generously given me both their time and shared their wealth of knowledge. This book does not necessarily reflect their exact point of view, but they have had a big influence on shaping mine. They include Joshua Lederberg, Philip Russell, C. J. Peters, Michael Ascher, Jerry Parker, Peter Jahrling, Ed Eitzen, Art Friedlander, Kevin Tonat, Frank Malinoski, Christopher Davis, William Patrick, Ken Alibek, Ben Liu, David Pui, and Scott Lillibridge.

Two of the very best authors in the business have been both sources of inspiration and information — Laurie Garrett and Richard Preston. They have provided me with many hours of captivating reading and thoughtful and engaging discussion.

One of the highlights of my professional career was advising His Majesty, the late King Hussein of Jordan. He is a giant that the whole world needs and desperately misses as it relates to our response to terrorism. Of course, without the kind and thoughtful assistance of Walt Wilson, his physician, I would have never had the opportunity to know such human greatness.

I've been fortunate to have colleagues who have unusual insight into the policy issue of bioterrorism — Janet Shoemaker and Peggy Hamburg. They have provided me a great deal of wisdom along with the facts.

Over the course of my career five individuals have been

there for me day after day. Kristine Moore, Michael Moen, Terry O'Brien, Phillip Peterson, and Jeffrey Davis have been, and continue to be, my professional compass. They have patiently provided me wise and sage advice on many occasions.

Thank you to Peggy Johnston for her twenty-five years of support. Most of all, I thank my wife, Barbara Colombo, and our children, Erin, Ryan, Andrew, and Maria. This book is really for and about you. Your support and love is an everyday gift that is priceless. I never want you to have to experience the hell of a bioterrorism attack. For you alone, every effort put into this book and all my other activities addressing this issue are worth it.

—Michael T. Osterholm
Greenwood, Minnesota
May 2000

Mike Osterholm made this book possible—he's a man on a mission, with the kind of enthusiasm and leadership that helped me find resources within myself to do things I might never have thought possible. Thanks are due as well to my agent, Rafe Sagalyn, for playing matchmaker and encouraging me to write this book. I owe everything to my wife, Jeanne Mixon, for her support for my decision to write this book and for her patience with me for the time the book stole from us. Thanks also to our wonderful kids, Elizabeth, Sam, and Joey. I wrote it for y'all; I want to see you grow up happy and healthy, in a world less threatened by the prospect of biological assault.

Just about every journalist I ever met had a passion for getting the story right. But as I learned more about this trade, I realized that many reporters believe that "getting it right" is solely about factual accuracy: making sure that the names are spelled right and that the bill actually passed by the number of votes you wrote down. That's fundamental—but getting the facts right is only half of the job. The other half of the equation is getting the tone right. That often means that when other reporters are trying to scare readers, it's up to you to explain why things might not be so very bad. During the Ebola scare of 1995, for instance, I wrote about how unlikely it was that the outbreak in Kikwit, Zaire, would spread to the United States, even though it was theoretically possible, and how hemorrhagic diseases have been contained in the past by standard Western health care practices. When many journalists were saying that the Y2K computer glitch was going to bring a kind of techno-apocalypse, I was among those who said that any problems that did emerge could be handled without much disruption. I have always looked for the center path in a story, the one between the doomsayers and the naysayers—so much so that I occasionally joke that my motto is "dare to be dull." Don't jump to conclusions, don't follow the herd, and most importantly don't scare people when the circumstances don't warrant the alarm.

The other side of that coin, however, is that when people are too complacent you *do* have to sound an alarm. In my discussion with Mike Osterholm and the people we interviewed and dealt with in putting this book together, I realized that this was a topic worth getting scared about. Not the kind of fright that leaves people feeling helpless, or gives the impression that nothing can be done. Just as the *Highlights* magazines of my childhood promised "fun with a purpose," you can think of this as "fear with a purpose"—scaring us all into taking action.

—John Schwartz
Takoma Park, Maryland
May 2000

If the device that exploded in 1993 under the World Trade Center had been nuclear, or had effectively dispersed a deadly pathogen, the resulting horror and chaos would have exceeded our ability to describe. Such an act of catastrophic terrorism would be a watershed event in American history. It could involve loss of life and property unprecedented in peacetime and undermine America's fundamental sense of security, as did the Soviet atomic bomb test in 1949. Like Pearl Harbor, this event would divide our past and our future into a before and after. The United States might respond with draconian measures, scaling back civil liberties, allowing wider surveillance of citizens, detention of suspects, and use of deadly force. More violence could follow, either further terrorist attacks or U.S. counterattacks. Belatedly, Americans would judge their leaders negligent for not addressing terrorism more urgently.

—Ashton Carter, John Deutch, Philip Zelikow,
"Catastrophic Terrorism: Tackling the New
Danger," *Foreign Affairs*

You have to be lucky all the time—we have to be lucky just once.

—Irish Republican Army

INTRODUCTION

It is a prospect so terrifying that the very thought of it can rob our world leaders of their sleep.

Explaining a 50 percent increase in monies proposed for preparing the nation for biological or chemical attack, President Bill Clinton told reporters from *The New York Times* that the "highly likely" threat of a terrorist group attempting such an attack on U.S. soil within the next few years "keeps me awake at night."

Me, too—and that's why I wrote this book. As an expert member of groups that have studied biological terrorism and an adviser to the U.S. government and others, I do not believe it is a question of *whether* a lone terrorist or terrorist group will use infectious disease agents to kill unsuspecting citizens; I'm convinced it's really just a question of when and where. The allure of the more deadly and frightening biological weapons has grown, and there's no shortage of people willing to kill for a cause—even a cause of their own imagining. What's more, the tools of bioterrorism are cheaper than ever before, and the skills to put it all together are within the reach of reasonably talented graduate students. Most important, the microscopic killers themselves are uncomfortably accessible, available by mail or on the Internet or within labs around the world.

Little wonder, then, that Dr. D. A. Henderson, the man who led the successful campaign to rid the world of smallpox, says biological terrorism "is more likely than ever before and more threatening than either explosives or chemicals." Now head of the Johns Hopkins University Center for Civilian Biodefense Studies, Henderson calls the specter of biological weapons use "every bit as grim and foreboding as that of a nuclear winter."

The American people are beginning to get the message. Nearly two-thirds of us anticipate a serious terrorist attack on the United States using chemical or biological weapons within the next fifty years, according to a 1999 survey by the Pew Research Center for the People and the Press. The same survey, oddly enough, found an overwhelming feeling (among 81 percent of adults) of general optimism about the future. They must assume that somebody else will get hit with the killer bug.

It's not that simple. As I will show, an attack with a highly contagious agent like smallpox could kill hundreds of thousands of people and could travel from city to city as easily as people do. Even a microbe like anthrax that doesn't spread from person to person could easily kill more than 100,000. We are, in other words, at the brink of a new age: what some experts call catastrophic terrorism.

Despite this chilling prospect—and the President's stated concern—the United States is not doing enough to prepare. Today's programs for combating terror by microbes are turning out to be more governmental public relations initiatives than workable policy. A classic Washington story, perhaps, but it's one with special urgency, given the dire results we could expect from an attack. Even if the current initiatives were run effectively at the international, federal, state, and local levels, we would still fall short of the facilities and therapies we need to protect private citizens around the world against biological agents like anthrax, smallpox, plague, Ebola virus, and other potential tools of a bioterrorist.

This book will tell you what that will mean when the inevitable happens and what we as a nation can and must do now to protect ourselves, our families, and the basic elements of our society that bioterrorism puts at risk.

You will also learn why the billions being spent to combat terrorism miss the mark when it comes to fighting microbial terror. Most antiterrorism programs lump bioterrorism in with what are commonly called WMD—weapons of mass destruction—including nuclear devices, bombs, and chemical weapons. But as you'll see, biological weapons are fundamentally different: they are the stealthiest tools of mass destruction ever developed. They can be released silently, with no sign or even smell to announce the act. They can take days or even weeks for the infections they cause to show themselves.

They first announce themselves with deceptive quiet, in emergency rooms and doctors' offices. That's the first line of response—not the military, not the police, and not the elite squads that have traditionally been the focus of antiterrorism programs. Early detection and action by these medical workers can create the opportunity for a strong response, with proper medication, vaccines, and even quarantine measures helping to mitigate the overwhelming damage that a bioterrorist attack will cause. A weak response, however, can amplify the damage and the crisis. Every case of a highly contagious agent like smallpox that is mistaken for plain flu, for example, sets up the potential for subsequent waves of infection. The health care, public health, and legal systems as we know them would quickly become overwhelmed.

This book shows why biological terrorism is becoming so much more likely, and examines the challenges of government response to bioterrorism. To illustrate the facts underlying these issues, each chapter begins with a portion of a fictional—but realistic—scenario describing what could happen in a bioterrorism attack involving (1) anthrax, (2) a foodborne pathogen, and finally (3) the nightmarish and highly contagious smallpox.

While writing this book, I have wrestled with the question of how much to tell. The result is hardly a how-to manual for terrorists. For one thing, the people who are serious about this already know more than we could ever say in this slim volume, and I have stopped short of giving away any secrets I've come across about the dangers we face and what lies ahead. There is nothing here to help individuals interested in perpetrating a bioterror hoax, either; if anything, I hope that this book will reduce the number of false scares out there by giving readers and the authorities the blueprint they need to respond more appropriately to bioterror hoaxes, thereby eliminating the thrill that the perpetrators of those hoaxes get by causing a panic.

Nor is the goal to push the panic button. John Schwartz and I—although the book is told from my point of view, the writing has been a full collaboration—want to warn you that the threat of biological terrorism is real without frightening you out of your wits. Instead, we hope to frighten you *into* your wits. The book you hold in your hands is a manifesto, urging each of us to prepare for the inevitable or pay a stupendous toll in human lives. It lays out a prescription for survival—for our nation and for us all.

—Michael T. Osterholm

1. TINY KILLERS

It's a good life for a germ in here. Out there in the real world, life is so uncertain: food sources can disappear, heat or cold can quickly reach deadly extremes, even a ray of sunlight can provide a fatal, incinerating blast. But here in this flask, it's very, very sweet. The temperature is just right, and food is all around. There's even some gadget in here that runs gentle streams of air bubbles through the brownish broth, making it a little foamy and ensuring that oxygen and nutrients are available for every single cell.

In such a supportive environment, the bugs do what they do best: consume nutrients and reproduce. Eat and multiply, eat *and multiply, over and over again. The tiny seed sample of* Bacillus anthracis *has repeatedly divided a few times each hour, growing in days into a rich soup of trillions of anthrax cells.*

The colony of rod-shaped bacteria is growing in a small microbiology lab, but not one at a drug company or university. The equipment sits on a couple of white Formica-covered tables in the storm cellar of a run-down old farmhouse; nutrients, stabilizers, and other chemicals sit on functional steel shelving against the whitewashed walls. The room gets its sterile greenish light from cheap fluorescent fixtures hanging overhead. The equipment—the thermostatic oven with its precise temperature controls, the heavy flasks and aeration systems, the desiccation and milling gear—would be familiar to scientists around the world. But their darker purpose would not be obvious.

Suddenly for the cells in the flask, the world becomes a less hospitable place. Taken from the cozy dark, they are poured onto a flat tray where warm air blows: a drying cabinet. The liquid all around is gradually going away.

The bacterium doesn't know many things, but it knows what to do when times get tough: it forms spores. First, its chromosomes fuse, forming a compact hunk of DNA. That DNA divides, and new molecular machines kick into action, creating a membrane to separate the two hunklets and adding an inner cortex and a series of protective outer layers. Finally, the mother cell bursts open, releasing its tough little pod. In that form, it will wait—for decades, if necessary—until it finds the next warm, wet place.

When dried and milled into lethal fineness, the resulting powder will be, ounce per ounce, more deadly than any explosive— and "smarter" than the most expertly programmed smart bomb. At this consistency, a handful of powder scattered into the open air dissipates immediately, each spore invisibly carried aloft by the undetectable currents of air we breathe. Even a very few spores, when inhaled, are enough to wreak chemical havoc on the body that unknowingly takes it in. Once lodged in the hospitably warm, wet environment so conducive to its growth, the spore will begin to thrive. As it grows in the body, it produces waste product that is a rich cocktail of deadly toxins. Like a killer who cuts the telephone line before entering the house, the poisons will move throughout the body and expertly slice up a protein called MAPKK. Cells use MAPKK as part of the chain of chemicals that carries some of life's most important signals, such as when cells should divide; without it, vital bodily functions simply shut down. In that way, the bacterium eventually kills the body whose warm, wet recesses nurtured its growth. Some of the people infected by the spores might be saved if they are lucky enough to be among the few who had been vaccinated, or if they get early antibiotic treatment before the toxins build up to fatal levels. But most of the people who breathe in the microscopic assassins will have no immunity, no protection—and no chance.

THE ANTHRAX BACTERIA in this scenario are taking part in an ancient war between humans and the pathogens that sicken and kill them. Our ongoing struggle against infectious diseases began with the first humans; we have vacillated between victory and defeat ever since. For much of the latter half of the twentieth century, humans appeared to gain the upper hand. But in recent years, diseases such as AIDS and the emergence of antibiotic-resistant bacteria, viruses, fungi, and parasites have reminded us that we remain deeply vulnerable to our tiny adversaries.

Throughout history, epidemics—those occurrences of illnesses beyond the expected—have demonstrated the potential danger of uncontrolled infectious diseases. As we enter the twenty-first century, we have developed enormous knowledge about the molecular structure of infectious agents and a cornucopia of treatments. But you can bet that widespread panic will occur in your community when several cases of bacterial meningitis show up among schoolchildren.

I've seen it firsthand, more often than I'd like to remember, in my two decades at the forefront of infectious disease epidemiology. Surprisingly, some of the most nervous people we've had to deal with have been health care workers. I can't count the number of times I've seen hospital staffers panic when a case of meningococcal meningitis came into a hospital. And that's a disease that isn't very contagious: the only way to get it is to come into contact with the patient's saliva. But many health care workers, the people most likely to know that simple fact, don't know it—or seem to forget it. I've seen 250 hospital workers demand antibiotic treatment just because somebody with meningitis passed through the building.

You'll see similar anxiety—or even outright panic—accompany the announcement that hamburger in your town might be contaminated with *Escherichia coli* O157:H7, that the neighbor's dog that every child plays with has just died of rabies, or that mosquitoes in your area might be carrying a virus that causes encephalitis. Yet by public health standards, or in terms of a germ's real potential to cause human disease, these examples are fairly routine. So what is it about infectious agents, those invisible germs we live with every day, that causes such a sharp human response?

Such respect and fear of infectious disease goes back to our ancestors—you could even say it's hard-wired into the psyche of our species. Plague, caused by the bacterium *Yersinia pestis*, can be traced to the earliest records. In the *Iliad*, Homer makes reference to a plaguelike illness common during the Trojan War (1190 B.C.), noting that it was associated with the movement of rats into populated areas. History has subsequently recorded three great plague pandemics (epidemics over large geographic areas). In the first pandemic, as many as 100 million Europeans, including 45 percent of the people of Constantinople, died of plague. The second great pandemic, known as the Black Death, started early in the fourteenth century, spreading across Europe and Asia. Between the years 1346 and 1352 alone, it claimed the lives of more than 24 million people— representing 25 percent of western Europe's total population at the time. Another 20 million Europeans had died by the end of the fourteenth century. Christians blamed the disease on Muslims, Muslims on Christians, and both Christians and Muslims blamed it on Jews or witches.

The third plague pandemic arose in China in 1894 and spread throughout the world via the beginnings of modern transportation. During that outbreak, it was discovered that *Yersinia pestis* caused the disease and that the primary reservoir for the disease was fleas on infected rats. In Manchuria plague caused an estimated fifty thousand deaths between 1910 and 1911. The pandemic arrived in Bombay in 1898, and over the next fifty years more than 13 million Indians died as a result.

It is against this historical backdrop of plague's horror that we can understand our very human fear of infectious diseases—the same kind of dread we associate with primal forces of nature like earthquakes and hurricanes. Optimists may argue that our century's sophisticated medical systems and improved sanitation have brought us to the happy moment when our worries regarding infectious diseases are over. But the potential for germs to kill millions is unfortunately still present today, even if the overall risk is reduced. That could be the most important lesson of the AIDS epidemic, which will not be the last emerging disease we will confront.

Even unwarranted fear of disease can explode in our modern world. Consider the panic in 1997 surrounding "Asian flu," the influenza epidemic in chickens and eighteen residents of Hong Kong. Many were convinced this episode would lead to another 1918 Spanish flu pandemic, the global tragedy in which more than 20 million people died. Even though there was no evidence of the influenza virus spreading from Hong Kong, the entire episode was front-page news and a lead TV story for several weeks. Travel to Hong Kong dried up overnight, and millions of healthy chickens were sacrificed to prevent the potential spread of the virus among them. The medical problems associated with that flu might have been dealt with effectively by the tools of modern medicine, but nothing in that arsenal could prevent the panic.

The fear remains. The next flesh-eating strep or food-borne disease episode, the identification of another young child who died of antibiotic-resistant *Staphylococci*, a possible case of mad cow disease in the United States, or an

Ebola virus epidemic in Africa will push all our ancestral buttons. Any of these events will automatically be major news stories and will constitute the obsessive talk of cocktail parties and church bingo games. This fear—hoary, deep, and ever present—is the backdrop for germs as weapons of bioterrorism.

GERMS AS WEAPONS

For a weapon to be considered effective and usable, it must have four main attributes. First, it has to be within the economic means of the user; a nuclear bomb may be the ultimate explosive device, but if a terrorist can't afford to buy or build one, it's not worth considering. Second, the weapon has to be capable of reaching the intended target; a missile that misses its target is useless; a gun that won't fire is a paperweight. Third, it has to cause limited collateral damage; a weapon that kills both sides creates a zero-sum game. Finally, the weapon has to result in the desired outcome, usually death.

No other weapon created by even the richest and most technologically sophisticated countries can match infectious agents in each of these categories. Unlike the large, expensive physical plant required to produce nuclear weapons' fissile material, or the elaborate refinery needed for a chemical weapons program, an effective biological weapons program can be set in a typical suburban basement, using basic high school or college lab equipment and materials easily ordered from catalogs.

Chemical weapons such as sarin and VX may be incredibly cheap compared to nuclear arms, but biological agents are cheaper still. Although figures would now have to be adjusted for inflation, one government analyst some years ago determined that $1,500 of nuclear killing power would set an anthrax assailant back by only a penny. And if the agent is highly contagious like smallpox, ongoing transmission would reduce the relative cost even further; from an economic point of view, a contagious bioweapon is a low-price bomb that rises in value every day.

Beyond the enormous ability of biologic agents to kill, three other features make them uniquely effective as terrorist weapons. First is the "footprint"—the size of a potential agent release. Then there is the ability of the contagious agents to be transmitted person-to-person following the initial release, spreading damage throughout the population and significantly prolonging the terrorist event. Finally, the delay between infection and the onset of disease—days, weeks, or even months—will compound the panic and terror; not only will the event continue to unfold over time, keeping the story fresh and hot in the news, but those who may have been exposed will be left with an ever-greater sense of dread, unable to know if they are already doomed to become part of the mounting death toll.

The visual image of the mushroom cloud rising above Hiroshima reminds us more than a half century later of the horrifying destructive force of nuclear weapons. It's an image that should never leave our consciousness. But next time you think of it, remember this: if efficiently delivered under the right meteorological conditions against an unprotected population, biological weapons will, pound for pound, exceed the killing power of any nuclear device.

Chemical weapons and nonnuclear bombs don't come close to matching the potential "footprint" of the biological weapon.

One of the touchstone documents in understanding the biological warfare threat is the 1993 congressional Office of Technology Assessment (OTA) report entitled "Proliferation of Weapons of Mass Destruction: Assessing the Risks," in which the federal government compared the hypothetical lethal impact of chemical, biological, and nuclear weapons aimed at the Washington, D.C., metropolitan area. Its shocking conclusion: a small airplane dispersing only 100 kg of anthrax spores—about 220 pounds—would be more lethal than a Scud-like missile carrying a hydrogen bomb. The OTA estimated that if the airplane dispersed its deadly cargo on a clear, calm night, 1 to 3 million people would die in a three-hundred-square-mile area surrounding Washington, D.C. The hydrogen bomb, by comparison, would kill 570,000 to 1.9 million individuals. (This study, which should have served as a global wake-up call for policy makers, remains little known today, although it's readily available on the Internet.)

Using an airplane to spray a metropolitan area is only one way to disseminate a biologic agent. An attack on a more compact scale using small spray devices could also deliver high concentrations of selected agents within the smaller footprint of public buildings or subways, and with devastating effect. Given the daily traffic in and through large buildings such as airports, office towers, and shopping malls, the number of persons exposed could equal the population of a big city. None of the technology required to deliver these agents from either airplanes or aerosol-generating devices is complicated or difficult to procure. Today, in our ever-evolving technological world, both methods have legitimate industrial applications—what experts in weapons of mass destruction call dual-use technologies.

In addition to the enormous potential footprint or initial size of the biologic agent release, some agents will be transmitted from infected persons to close contacts. This is in effect making a bomb that continues to explode—and, depending on the agent used, the result could be truly catastrophic. Smallpox represents, perhaps, the worst possible case in this regard. The virus is so highly contagious that it can be passed in the air to a person in the same room as the victim; it is hardy enough to remain stable and infectious in the environment—suspended, say, in a large room or the cabin of an airplane—for long periods of time. Imagine AIDS or Ebola transmission through casual contact at the office, school, or even a shared elevator ride. If the smallpox virus were to emerge anew, even a limited number of cases would soon result in uncontrollable second, third, and even fourth waves of infection and death.

Now take that imagined horror a step further and see yourself being told that you were likely exposed to an infectious agent in the air and you might get sick and die. It could happen tomorrow or next week or even next month. But no one knows for sure. Along with this agonizing fact, add another factor: the public health and health care systems likely won't have adequate supplies of the appropriate vaccine or antibiotic to help prevent you from becoming ill.

That's the terror of a biologic weapon. With a bomb or even chemical release, the impact of the assault is almost instantaneous. The climb back to normal life begins within minutes of the explosion or release. With a biologic agent like anthrax, however, the incubation period (the time between release and first onset of illness) may be forty to fifty days. The ability to identify the site of release, those exposed, and possible perpetrators is severely limited. After all, what if Timothy McVeigh had three weeks to get away from Oklahoma City before anyone knew what he might have done? What if the World Trade Center bombers had been able to set off their bomb without a blast, a flash, or even a sound? There would be no destroyed truck with serial numbers that could be used to track the killer, no defining event to send law enforcement officials into action. Even if it is known that a release of biologic agents occurred, the potential victims might not know if they were even exposed. Do you recall with perfect accuracy where you were three weeks ago at 3:18 P.M.? This inability to rapidly identify a crime scene and commence treatment of those exposed, as well as deal with the possibility of illness with an unknown agent, adds to the potential for terror beyond the deployment of any other weapon.

AN IDEAL BIOLOGICAL TERRORISM AGENT

The world is home to thousands of known infectious agents or pathogens, which cause disease in humans, animals, and plants. A smaller, yet still large number of viruses, bacteria, fungi, and parasites cause damage or intoxication in humans. But only a very limited number of agents will be effective in weapon form. For a biological agent to create both disease and terror in a civilian population, it must be easily procured and easily produced. You'll see later how easy it is to obtain the biological agents likely to be used in an attack against civilians. Once in hand, anyone with a basic background in microbiology and some moderately priced equipment can easily prepare the quantities of agent needed to cause significant illness in a population. And as Iraqis and former Soviets have demonstrated, many agents can be produced by the ton. The most powerful agents are capable of being dispersed in aerosol form, in particles of one to ten microns in size. This particle size is able to drift far enough to reach people miles away and still be small enough to be taken in by the lungs. While airborne, the pathogen must be able to survive sunlight, drying, and heating until it is inhaled; after all, agents surviving for hours can expose many more people to disease than those that die after a few minutes. Similarly, foodborne and waterborne pathogens, which are generally less lethal, must survive the holding temperature and pH levels of the food or water they are placed in.

Detection, panic, and terror will of course be limited to those agents that cause a lethal or disabling disease. An epidemic of common coldlike symptoms, if noticed at all, won't disrupt our way of life or push our ancestral buttons. If the deadly agent is highly contagious, however, the terror will grow with each new wave of infection.

If there is any comfort a civilian population will seek during an attack, it's the saving power of high-tech medicine — the notion that we can pop a pill and everything will be all right. There are a number of agents for which there is an effective treatment or preventive therapy for those who have been exposed but who have not yet fallen ill—treatments that doctors call prophylaxis. But of course, if adequate supplies of those medicines aren't readily available, the ensuing terror will spread almost like a second highly infectious assault. Fortunately, this factor is also one we can affect by developing needed vaccines and preventive treatments.

Experts who have considered the issues have come up with various lists of the "threat agents." The NATO handbook dealing with potential warfare agents lists thirty-one pathogens. Only a limited number, however, can be grown and dispersed effectively to cause a large number of serious illnesses. In 1994 a group of former Russian biological warfare experts disclosed that they had worked with eleven such agents. The four highest on the list were smallpox, plague, anthrax, and botulism. But the Russians admitted that they had also developed tularemia, glanders, typhus, Q fever, Venezuelan equine encephalitis, and the Ebola-like hemorrhagic killer known as Marburg virus.

In 1998 I was part of a working group organized by the Johns Hopkins Center for Civilian Biodefense Studies, which considered potential agents that present the greatest risk for transmission, ability to infect large numbers of civilians, and, of course, cause death. We concluded that smallpox and anthrax had the greatest potential for mass casualties and civil disruption. Others of serious concern were plague, botulism, tularemia, and viral hemorrhagic fever.

In June 1999 I served on a similarly focused CDC panel of experts in medicine, public health, military intelligence, and law enforcement. We first evaluated biological agents according to four criteria: the public health impact based on the potential for disease and death and the ease of transmission; delivery potential based on the ease of mass production and distribution, as well as stability of the agent; special preparation requirements based on the availability of antibiotics, antiviral drugs, immune globulin, and vaccines; and finally public perception of the agent and its ability to generate fear. We then ranked the agents in three categories according to a risk matrix analysis. Category A contains agents considered the greatest threat in terms of producing casualties and in terms of the need for stockpiling antibiotics and vaccines. Category B agents were considered to have potential for transmission and illness but with fewer requirements for public health action. Category C agents were those considered as possible emerging public health threats. Our findings are shown in the table below.

BIOLOGICAL AGENT	DISEASE
CATEGORY A	
variola major	smallpox
Bacillus anthracis	anthrax
Yersinia pestis	plague
botulinum toxin (*Clostridium botulinum*)	botulism
Francisella tularensis	tularemia
filoviruses/arenaviruses	hemorrhagic fever

CATEGORY B

Coxiella burnetii	Q fever
Brucella species	brucellosis
Burkholderia mallei	glanders
Burkholderia pseudomallei	melioidosis
alphaviruses	viral encephalitis
Rickettsia prowazekii	typhus
toxins (ricin, Epsilon, toxin of *Clostridium, perfringens*, staphylococcal enterotoxin B)	toxin syndrome
Chlamydia psittaci	psittacosis
foodborne disease agents	
waterborne disease agents	

CATEGORY C

Any agent identified by the CDC Emerging Infectious Diseases program, such as Nipah virus, hantavirus, and tick-borne hemorrhagic fevers

Source: Centers for Disease Control and Prevention, Critical Biological Agents for Public Health Preparedness, 1999.

LIKELY AGENTS OF BIOLOGICAL TERRORISM— THE "CATEGORY A" BUGS

Smallpox

Smallpox, the nightmare to end all nightmares that was eliminated as a natural disease in the 1970s, often starts with a simple fever—the sort of thing anyone might get. The disease has a relatively long incubation period (the time from exposure to first symptoms). If you were unlucky enough to be exposed to the virus, it would take about twelve days for those first signs of fever, headache, and malaise to appear. Next a red rash will appear in your mouth, throat, and on the face; soon after that it appears on the arms, legs, and torso. At first the rash doesn't look like much and can be mistaken for chickenpox. A trained health care worker could detect important differences, but it takes a special kind of person—the kind who, when the sound of galloping horses nears, considers that it might possibly be a herd of zebras.

Soon the horror begins. The pocks—bulletlike, pus-filled blisters—begin to appear on the skin. As you enter this stage, you're likely in agony; people have compared it to having their skin on fire. At this point, you've been highly infectious for several days, and for another week your body will keep pumping out virus particles with every breath, infecting your loved ones, your doctors, and anyone else unfortunate enough to come near. Only after several days of rash and the appearance of pocks is a physician likely to consider smallpox. Even then, that will probably only happen if the doctor is old enough to have seen it or has had sufficient training in biological agents.

The disease is caused by the variola virus. It is a naturally occurring disease that spreads from person to person through airborne respiratory droplets, or aerosols. If you're thinking you're immune, you're probably mistaken. Vaccinations ceased in the 1970s, so today's under-thirty population has no immunity; what's worse, many of the people who were vaccinated thirty or more years ago have lost or reduced their immunity as their vaccinations wore off. That leaves you unprotected, or with limited protection, against a disease that kills one in three victims. And if you do recover, you'll bear the disfiguring pockmark scars for the rest of your days. There is no cure for smallpox; although the chance of survival in a limited number of cases could possibly be improved by a currently available antiviral drug, it is difficult to use and in short supply.

Few people in the world can even remember smallpox: it's been more than fifty years since the last case occurred in the United States, and the last naturally occurring case in the world was reported in Somalia in October 1977. Since that time, it has been assumed that mankind's worst scourge was over. However, with the revelations that Russia, Iraq, and North Korea harbor undeclared stocks of smallpox virus, we must once again consider the possibility of its return.

To put smallpox into perspective, one only need look at what disease and death it caused throughout the world in the twentieth century. The world's population was much smaller, of course, through most of this century — 1.6 billion people in 1900, compared to 6 billion in 1999 — and smallpox was substantially reduced in most of the developed world by the 1940s. Still, despite its relative rarity through much of the century in so many areas of the world, approximately 500 million people died of smallpox in the century that just ended. This compares with 320 million deaths during the same period as a result of all military and civilian casualties of war, cases of Spanish flu during the ruinous 1918 pandemic, and all cases of AIDS worldwide. These staggering numbers make painfully clear how grave a global crisis any return of smallpox would represent;

the use of it as a weapon would constitute the ultimate crime against humanity.

If a case of smallpox were actually identified, everyone who had come in contact with that patient would have to be vaccinated immediately to reduce the risk of additional cases. But as we will see in later chapters, the supplies of vaccine have dwindled to 15.4 million doses in the United States inventory, which is held by the CDC. Of note, the CDC has come up with the 15.4 million estimate; other experts seriously doubt if we could get 7 to 8 million doses out of our current stockpile. Either number may seem like a lot of doses, but the number is deceptive: during a relatively small 1972 outbreak in Yugoslavia, that nation of 21 million people required 18 million doses of the vaccine in only ten days.

Currently, the smallpox vaccine in our inventory as well as around the world is primarily derived from the old process of scarification of calves—scratching the virus vaccinia (not smallpox, but cowpox) into the skin of a calf and harvesting the subsequent infection. Most of our severely limited vaccine supply is more than twenty-five years old. For safety reasons, we can no longer use a live cowpox vaccine, which is essentially "cleaned up" cow pus. Even though in early 2000 the federal government has finally initiated activities to develop and produce a new vaccine from cell cultures, it will be years before new and sufficient vaccine is available.

If an actual release of smallpox (whether intentionally or by accident) were to lead to an epidemic, only early detection, isolation of the infected individuals, surveillance of their contacts, and a focused selective vaccination program will allow us to regain control.

Anthrax

Anthrax is a brilliantly efficient killer. Most people who come by the disease naturally have been exposed to infected animals—eating contaminated meat or handling the skins or wool of infected animals. The bacteria can enter the body through ingestion—that is, eating the contaminated meat—or through contact with broken skin, which develops into what is called wool-sorter's disease. When it enters through the skin, antibiotics are often sufficient to bring about a full recovery.

Breathing the bacteria into your lungs, however, causes a different form of the disease: inhalational anthrax. This is the most deadly form, and the most likely one that a terrorist would try to exploit. If you breathed the spores into your lungs, you would probably be ill within two to ten days—but your body could hold off showing signs of illness for the next six to seven weeks. When it does hit you, it's swift and ruthless. As the bacteria grow in the lymph nodes of your chest, early symptoms mimic many common flulike illnesses. By the time you've got a full-blown case and get a proper diagnosis, antibiotics and intensive medical care are unlikely to help. If you're like most patients, you'll be dead within twenty-four to seventy-two hours from overwhelming infection and shock caused by toxins that the bacterium produces.

There is substantial evidence that antibiotic treatment and use of anthrax vaccine after exposure but before symptoms can greatly reduce both the number of illnesses and the number of deaths. This creates other problems, however, since a bioterrorism release will likely involve far more people than our current supplies of both could help. If there is any good news about anthrax at all, it's that there is no evidence that the disease can be passed from person to person. Still, *Bacillus anthracis*, the agent that causes anthrax, is the most serious of the bacterial threats that we face because it is the likeliest one to be used. Easier to get and grow than smallpox, it is the biological agent that the Soviets, the Iraqis, and even the United States (in the 1960s) gave a high priority to for "weaponization," or turning into a weapon.

Plague

As its historical name implies, plague is a disease that has evoked panic and fear in populations dating back to our earliest history. People usually contract the disease, caused by *Yersinia pestis*, after being bitten by an infected flea. The bite leads to bubonic plague, with distinctly swollen and painful lymph nodes—the "buboes" that give the disease its name. Appropriate and timely treatment with antibiotics can cure this form of the disease. Those who have developed biological weapons, however, have generally chosen a more direct and potent route: inhaled aerosol.

One to six days after you breathe in the microbe, you are likely to develop the especially virulent form, pneumonic plague. If you don't get treatment, you will quickly slide into kidney and respiratory failure and subsequent shock followed most likely by death. Although there is no available vaccine for plague, rapid use of antibiotics for those likely to have been exposed to the aerosol of plague or to cases of pneumonic plague is crucial. As with anthrax, the difficulty of getting adequate supplies of antibiotics to the victims in time to help them is daunting. Unlike anthrax, however, plague has been shown to pass from person to person. It's not nearly as contagious as smallpox, but an attack would certainly affect more people than those caught up in the initial exposure.

Botulinum Toxin

Botulism is typically a foodborne ailment caused by the toxins of *Clostridium botulinum*. Ingesting the tiniest amounts usually leads, twenty-four to thirty-six hours later, to blurred vision and difficulty in swallowing and speaking. Depending on the severity of exposure — and ingesting minuscule amounts of the toxin can have dire results — the symptoms may progress to general weakness, respiratory failure, and death.

In a large outbreak, all you can hope for is that your community has enough respirators available to keep you alive when your own useless muscles can't do the breathing for you. Of course, that's not the case in any city in the United States today.

Tularemia

There are a few forms of the bacterial disease tularemia, but the only one likely to be used as a weapon is the inhaled, or typhoidal, variety. It causes you to develop fever, chills, headache, and general weakness, as well as chest pain, weight loss, and a nonproductive cough. It kills an estimated 35 percent of its victims. The effects come on quite quickly, usually within three to five days. An especially hardy bug—it resists freezing and can remain viable for weeks in water—tularemia was cultivated for weapons use by the United States in the 1950s and 1960s, by the Soviet Union, and by other nations as well. It can be treated with certain antibiotics. However, as with the case of an outbreak of anthrax or plague, if you came down with tularemia you would desperately need antibiotics—and supplies of those drugs aren't adequate to ward off the effects of a large-scale attack.

Hemorrhagic Fevers

Hemorrhagic fevers are the horrifying diseases described in books like Richard Preston's *The Hot Zone:* ailments that cause the body's fluids to leak out of tissues and orifices and cause particularly gruesome deaths. These viruses kill an estimated 30 to 90 percent of their victims. The most famous of these, the filovirus Ebola, has been known since 1976 but has never been seen in the United States. Its effects come on fairly quickly, usually within three to five days. It can be treated with some antiviral drugs—which, however, like key antibiotics, aren't available in large amounts.

Members of the Aum Shinrikyo cult flew to Zaire in hopes of finding samples to turn into biological weapons. They apparently were unsuccessful in finding the elusive virus. Another viral hemorrhagic fever, Marburg, has been associated with an outbreak in Germany and Yugoslavia that infected thirty-one people. According to former Soviet bioweapons official Kenneth Alibek, the Soviets tested Marburg extensively and based their weapon version on a strain that killed a scientist in a laboratory accident. The category also includes yellow fever, Lassa fever, dengue, and more. Some of these diseases respond to antiviral drugs, but these drugs are in short supply and would not be available for large populations of victims.

2. THE
INVISIBLE MAN

The head of the underground anthrax lab and sole occupant of the old farmhouse is Ed, a heavyset man in his thirties. Though grim by nature, he is at his happiest in this home-built sanctum. Other people's labs irk him; nobody ever seems to do it right. Back in the Army Medical Corps, the labs seemed sloppy and the work was always too rushed. He owed the military a debt of gratitude, he supposed, for giving him the foundation of his knowledge of microbiology, but the tasks had been dull and the people he worked with didn't seem to care. Ed could not abide people who did not take their science seriously.

Ed had hoped things would work out when he got back

home after his stint was up, but he just found himself in what he saw as another dead end. He was back home in the college town, living in the house his parents had left him and working as an assistant in one of the microbiology labs on campus. He would never make it through the door as a graduate student; the little breakdown he'd had during his undergraduate days had pretty much whacked his grades for good. So he got a job there instead, and hated it. Even more, he came to hate the professor who ruled it like some little fiefdom, who treated Ed like a lower life-form. Hell, the guy didn't even seem to remember Ed's name—even though it was Ed who always seemed to be able to make the difficult cultures grow, Ed who always figured out the ingenious way to rig up equipment for just about any experiment. Ed with the good hands, but no credentials. Ed, the genius looked down on by all of the educated fools.

It was during those frustrating years that he came up with The Plan. The hardest part, in retrospect, was selling the big old house. But he could never have built what he needed to build and kept it a secret in town. And the real estate market was so hot that the sale gave him more than enough cash to set things in motion.

Then came the farmhouse, out in the middle of nowhere. The farther away from other people, the better—not just because Ed was not exactly sociable but also to make sure that if he accidentally released his pets, they wouldn't kill the neighbors and draw a lot of attention. And the cellar was perfect.

Once he had moved into the place, turning that big old room into a lab was fairly straightforward. The previous owner had put in a good cement floor, and Ed scrubbed and painted the thick stone walls. He was as proud of that little basement lab *as the professor was of his big setup over in the ivory tower. The underground space was always cool, so he didn't need to worry about air-conditioning. His dehumidifier kept just the right amount of moisture in the air. But he did put in a good air filter to keep the place dust-free and handle exhaust air to prevent any accidental escapes. The equipment was easy to scrounge from storage rooms at the university; they were always buying the latest toys for themselves and never paid attention to what happened to the cast-offs. And the older stuff did everything Ed needed it to do.*

In just about every way, the lab was invisible to the outside world. Ed knew that the police sometimes look for indoor, grow-light marijuana farms by checking out utility bills, but he was confident that his kilowatts wouldn't betray him. He'd had to run a new electrical circuit into the cellar to handle all of the plugs, but even when everything was humming, Ed's place didn't draw as much power as some of the houses down the road. He'd even made enough space for a little typing table that had just enough room on it for a few microbiology texts and his laptop, which he used to surf the Net's multitudinous science sites. There was plenty of great information out there, and all free.

For a while, he tried to figure out a way to get vaccinated against anthrax, since he'd gotten out of the military a few years before they started vaccinating everybody. But he didn't want anybody asking why he needed it. So now he's careful. Always wears his mask and latex gloves when he's working. Always keeps plenty of antibiotics around and takes a series whenever he's doing the tricky stuff. Like today.

Benchwork has put Ed in a pleasant mood. He hums tunelessly to himself as he adds a chemical treatment he developed for the university lab that will keep the natural static charges on the spores from making them clump together—another process that he never got credit for! Once that's mixed in, he removes the aeration tube from a flask and pours the liquid onto a tray, where it forms a thin layer. He checks the filter over the exhaust pipe and then flips a switch on the cabinet, which will speed the process of evaporation. He's gone through the whole process dozens of times now, building his stockpile bit by bit. He could have saved time and steps with bigger equipment, but that was exactly the kind of stuff that would attract attention from people he doesn't want to talk to.

After setting up the drying process, Ed methodically shuts down the lab, cleaning all of the glassware and surfaces and putting his materials away. He throws away the mask and gloves and casually strips naked. He puts his clothes in the small washing machine by the lab bench, dumps in detergent and bleach, and starts it up. He then steps into the small shower and scrubs himself thoroughly, making sure that no spores will hitchhike out of the cellar on his body.

A few days later Ed drives his pickup truck to a little rural airstrip. It serves crop dusters and small-plane hobbyists for the most part; captains of industry weren't bringing their Gulfstreams into this dinky town. The grass runway shows that it's the sort of place for pilots who aren't picky—and who fly forgiving planes. No one is even around to notice him that day.

He gets to work quickly, stepping through the practiced rituals of getting his plane ready to fly. He doted on the little aircraft, his sole indulgence once he had conceived of The Plan. He'd had to wrestle with his conscience a little at first, since a committed warrior should maintain certain ascetic ideals. But he had found it for sale second- or fourth-hand on a Website for pilots and was able to pick it up without busting the budget he'd set for himself after putting The Plan in motion.

As he performs the flight check, Ed takes special pleasure in reexamining the crop-dusting equipment. He once thought that he was going to have to build the dispersal rig himself, but had been excited to find a dry pesticide unit; it saved him countless hours of trial and error. Hardly anybody literally "dusts" crops with pesticides anymore, since the powders spread so much farther than the target field. Nowadays everybody sprays liquid pesticides, which are easier to control. But what others see as a problem, Ed sees as a desired feature—he wants this stuff to spread for miles and miles. "Keep your powder dry," he thinks, and allows himself a chuckle.

The sun is almost setting by the time he's ready for takeoff. The engine starts with an angry roar and he taxies down the strip, pulling up into the clear evening sky with a feeling of exhilaration. "It's finally happening," he thinks to himself.

The evening comes on clear, cool, and dry; a light breeze from the west lifts the stillness just so. A full moon is rising and will soon light the fields with a silvery glow. It's the kind of weather that makes pilots happy to be alive.

As he flies along, Ed guides himself by his handheld global positioning satellite box, making his way in the deepening darkness to his target. The lights of the city come into view. He pulls up and circles around to approach the stadium from the west; 74,000 spectators in the stands below are watching an early-season game. As he nears the stadium, he turns once more and comes in low. Ed flips a switch that starts the duster. The anthrax spores that he has so carefully amassed stream out of the back end of the plane in a fine haze that vanishes as the particles disperse. He looks behind the plane with satisfaction. The fact that he can't see anything in his wake means there's no clumping; all of the care he put into prep work and milling has paid off. The breeze carries the spores toward the unknowing throng to the east, and beyond. He knows that much of the invisible dust will drift far away from the stadium, hitting many more downwind neighbors as well as the sports fans. It's not his goal, but there isn't much he can do about that. Drift is drift.

The hopper empty, Ed turns the responsive craft around and heads back toward the little airstrip. By now the invisible spores will have settled on the stadium crowd and others throughout the city; they are drawing the powdered death deep into their lungs by the thousands. His tiny assassins are on the job.

WHAT SORT OF person would *do* such a thing? Are there really people who could so carefully plan the deaths of thousands of people?

Plenty. Many of this new breed of terrorists don't simply kill innocents, they *intend* to—at American embassies in Tanzania and Kenya, at the Alfred P. Murrah Federal Building in Oklahoma City, and at New York's World Trade Center. According to evidence in one of the trials following the 1993 World Trade Center bombing, the conspirators actually intended to topple the north tower into the south tower, "just like a pair of dominoes," in the words of U.S. District Judge Kevin Duffy. The jury heard that the group also tried unsuccessfully to create a cyanide cloud in the explosion that would have killed still more.

Conspirator Ramzi Ahmed Yousef told the Secret Service agent who accompanied him on the flight from Pakistan to the United States after his arrest that he had hoped to kill 250,000 people in the midday attack.

People like these—people like "Ed"—are much on the minds of America's law enforcement, intelligence, and policy officials. U.S. Defense Secretary William Cohen might have been describing Ed when he warned in a July 1999 *Washington Post* essay of the "looming" chance "that these terror weapons will find their way into the hands of individuals and independent groups—fanatical terrorists and religious zealots beyond our borders, brooding loners and self-proclaimed apocalyptic prophets at home."

Terrorism itself is not new—as long as there have been nations, there have been people fighting to destabilize them and spread fear through stealth and ruinous attack. When John Wilkes Booth shot President Abraham Lincoln at Ford's Theater and shouted *"Sic semper tyrannis!"* as he fled, he was committing a terrorist act. Visionary novelist H. G. Wells illustrated the "terror" in terrorism with uncanny clarity in his 1897 novel *The Invisible Man:* the mad killer of the title threatens "a reign of terror" based on his ability to commit crimes unseen. The Invisible Man gloats that he will begin his attack on society with a murder; of his victim, he says, "He may lock himself away, hide himself away, put on armour if he likes; Death, the unseen Death, is coming."

Even though it has always been with us, terrorism is changing—morphing into something that law enforcement officials and policy makers can't even define well, let alone counter effectively. The FBI admitted in the 1997 edition of its annual "Terrorism in the United States" report that "there is no single, universally accepted definition of terrorism." The agency then provided the official, and very broad, definition it uses. "The U.S. Code and the FBI define terrorism as '. . . the unlawful use of force or violence against persons or property to intimidate or coerce a government, the civilian population, or any segment thereof, in furtherance of political or social objectives.' "

None of the advances in the biological sciences that put deadly tools within the reach of terrorists would mean anything if not for fundamental changes in the nature of terrorism alluded to by Cohen—changes that make today's bad guys both harder to track and likelier to use weapons of truly catastrophic terrorism than any who have come before them.

In the past—absurdly, what might be thought of as "the good old days" of terrorism—several factors made the world a more orderly place. The Cold War, for all its faults, created an overall geopolitical structure in which two superpowers faced off; the two sides exerted a measure of control over the terrorist-supporting nations and extremist groups they sponsored. Many of the terrorist groups of the time were tightly organized, which made it possible for law enforcement agents to monitor and even infiltrate them to a degree. What's more, the groups tended to worry about retaliation, and even about their image. The Irish Republican Army would commonly phone in warnings about its bombs so that innocent people could be moved out of harm's way. The old conventional wisdom on such groups was best summed up in an often-quoted line by terrorism expert Brian Jenkins of Kroll Associates, the international security firm: they want "a lot of people watching, not a lot of people dead."

Although some form of terrorism has always been with us, the face of terrorism has changed rapidly in recent years—and not for the better. Modern terrorist groups tend to be decentralized, and many self-declared terrorists work alone. Those motivations may include Islamic fundamentalism, white-supremacist hatemongering, end-of-the-world beliefs tied to the coming of a new millennium, or a pastiche of ideas that come together simply to validate a killer's violent urges.

And, increasingly, these killers will consider using weapons that can do catastrophic harm. Of course, most terrorist groups will continue to rely on conventional weapons and conventional numbers of casualties, since bombs and guns are simply easier to procure than anthrax or sarin. But Mike Reynolds, who tracks hate groups for the Southern Poverty Law Center in Montgomery, Alabama, explains that today's terrorists may be more likely to use cataclysmic weapons such as biological or chemical agents than classic groups like a national liberation movement. The liberation rebels, he says, hope to run their nation after the battle is over and would be afraid of turning their people against them with a horrific act. Circumstances are different, Reynolds says, if a group is not hoping to govern, but simply to destabilize and destroy—if "what you're looking for is to wipe out the infidel, the Great Satan, the Jews [and not] looking at any great result like sharing the table at the U.N. . . . You simply want to wipe them out."

With so many troubling trends coming together, "the

danger of weapons of mass destruction being used against America and its allies is greater now than at any time since the Cuban missile crisis of 1962." Those aren't the words of professional scaremongers: they're from a landmark 1998 article about the new "catastrophic terrorism," in the journal *Foreign Affairs*. One of the authors of the article is former CIA director John M. Deutch. (The other two are Ashton Carter, a Harvard University professor and former top official of the Department of Defense on international security matters, and Philip Zelikow, a Harvard professor of public policy.) In chilling 1999 congressional testimony, CIA official John A. Lauder said that at least a dozen terrorist groups have expressed an interest in or have actively sought nuclear, chemical, or biological weapons capabilities.

Their unique capacity for gradual destruction already attractive to any group attempting to sow the maximum amount of fear and chaos, biological weapons are becoming an important part of that armamentarium. Jessica Stern, author of *The Ultimate Terrorists*, warned in a recent paper in the CDC's journal *Emerging Infectious Diseases* that three trends are boosting the chances for bioterrorism. First, more groups are willing to take the political risk of causing widespread destruction; second, the biological agents and devices are becoming more readily available; and third, the groups themselves have the kind of decentralized structure that helps them to operate in secrecy. "The intersection of these sets is small but growing, especially for low-technology attacks such as contaminating food or disseminating biological agents in an enclosed space," she wrote. "Major attacks are also becoming more likely."

Understanding why terrorism has undergone this transformation requires a deeper look at several aspects of the phenomenon. It takes us through the risk still associated with the remnants of state-sponsored terrorism, with international and domestic terrorist groups, with the rise of destructive cults, and with an emerging trend of "lone wolf" attackers.

STATE-SPONSORED BIOLOGICAL WEAPONS PROGRAMS

It's one of those moments in life that we've all experienced. Time comes to a standstill. You can remember where you were, what you were doing, whom you were with, and maybe even the clothes you were wearing. Depending on one's age, it may have been the moment that radios throughout the United States announced the attack on Pearl Harbor as it was unfolding. It may be more recent, when the weekend morning news bulletins announced the tragic death of Princess Diana or the disappearance of the plane piloted by John Kennedy, Jr.

For me, the first such moment was that November afternoon when Sister Clarice walked into my fifth grade classroom to announce that President Kennedy had just been shot and killed in Dallas. Another moment that following Sunday was also burned into my memory: I sat in the old blue Chevy Impala, with my uncle Matt and my cousins, listening to the radio announcer scream that Lee Harvey Oswald had just been shot only a few feet away from him.

Those moments stay with me, and I think of them now and then. But one moment comes back to me every day. It occurred on May 11, 1993, at 1:05 P.M. eastern daylight time, at 1600 Clifton Road in Atlanta, Georgia. I sat on the north side of Auditorium A at the Centers for Disease Control and Prevention (CDC). As a member of the National Centers for Infectious Diseases Board of Scientific Counselors, I was attending our biannual meeting. Years later I still remember that afternoon with startling clarity: the faces, the seating arrangement, and the feeling that turned from interest to concern and then to fear. The board was meeting in closed session that afternoon; members of the public were not allowed to attend. It's usually uneventful.

Our charge that afternoon was to provide guidance to the CDC on the destruction of the smallpox virus. A debate was emerging in the Clinton administration about whether we should postpone the scheduled destruction of the last-known smallpox virus cultures. Almost everyone in the public health and medical world agrees that the greatest accomplishment of the collective human effort to reduce pain and suffering was our victory over smallpox. The campaign, begun in 1967 under the aegis of the World Health Organization (WHO), succeeded in eradicating the horrible disease in 1977. In 1980 the World Health Assembly recommended that all countries shut down their vaccination programs. When a WHO expert committee recommended that all laboratories destroy their stocks of smallpox virus or transfer them to one of two WHO reference laboratories—the Institute of Virus Preparations in Moscow or the CDC in Atlanta—we all naïvely believed those recommendations would be carried out.

What made that day in 1993 so viscerally memorable was a briefing on information provided by two senior defectors from the former Soviet Union's bioweapons program. Their message: smallpox wasn't safely locked up.

The Soviet scientists reported that beginning in 1980, they had embarked on a massive program to produce the smallpox virus in quantities beyond human comprehension, and to adapt it for use in bombs and intercontinental ballistic missiles. The program had an industrial capacity capable of producing many tons of smallpox virus annually. With the demise of the Soviet Union, it is possible and even likely that these weapons—or even just a test tube of the smallpox virus—would find their way into the hands of those who would reintroduce it to a now very susceptible global population.

Next to me at the meeting that day was retired general Philip Russell, M.D. P.K., as he is known by his friends, had a distinguished career in the area of infectious diseases and had served as the commander at the army's Medical Research and Development Command. Many people have come to know Phil from the portrayal of him as a key decision maker in Richard Preston's book *The Hot Zone*. We whispered back and forth throughout the meeting, concluding that we were in a horrible mess. That evening, after dinner, several drinks, and lots of additional discussion, the future looked even more bleak. We realized that with the very limited availability of smallpox vaccine in the United States—and, for that matter, throughout the world—even an accidental release of this virus into civilian populations would result in catastrophe.

The Soviets were not alone. Many states still support terrorism and many are manufacturing catastrophic weapons for battlefield or terrorist use. The United States has identified twenty-five countries that have or are developing nuclear, chemical, and biological weapons and ballistic missile systems to deliver them. According to a 1999 report by the congressionally created Commission to Assess the Organization of the Federal Government to Combat the Proliferation of Weapons of Mass Destruction, most of the nations identified as sponsors of terrorism either have or are seeking weapons of mass destruction. (Those nations are Cuba, Iran, Iraq, Libya, North Korea, Sudan, and Syria.) According to the commission, chaired by Deutch, more than a dozen states have offensive chemical and/or biological weapons programs. And in a recently unclassified CIA report on proliferation of weapons of mass destruction, the agency singled out Iran, Iraq, North Korea, and Sudan as worrisome nations in the area of biological weapons.

Americans who wonder where these nations' bioweaponeers get their technology might not need to look any farther than their own borders. The United States has helped to educate generations of fighters and scientists whose expertise is now being used against us and the rest of the world. Just as the military training that the United States lavished on Afghan mujahideen serves the extremist Taliban forces well, the technological hurdles that need to be jumped by potential bioterrorists have been lowered greatly by America's open access to education.

Of course, the vast majority of those trained in these institutions intend to use their knowledge for benign purposes. But not all of them. Stephen Morse, a respected colleague currently serving as an official in the bioterrorism response offices of the CDC in Atlanta, told me that he received a call from American intelligence agents not too long ago about Ahmad R. Bahrmand. Morse immediately recalled Bahrmand, an Iranian graduate student who worked in Morse's lab in the mid-1980s. At the time, Morse was heading a lab researching sexually transmitted diseases, and the Iranian student came to work for him as a researcher and technician, learning the basic skills of microbiology at one of the nation's premier scientific institutions.

Today, the agents informed Morse, Bahrmand heads Iran's biological warfare program.

Morse draws some comfort from the fact that he felt that Bahrmand was relatively incompetent. "I had to fire the guy . . . if this is an example of someone running the Iranian BW program, we've got nothing to worry about." Scientifically, Bahrmand was a sloppy worker who consistently contaminated his samples—in some cases, by sucking fluids into pipettes by mouth instead of using sterile bulbs. Morse recalls that Bahrmand bragged that he was connected to the family of the Shah of Iran—a claim that turned out not to be true. Most troubling for the soft-spoken Morse, Bahrmand showed up on his doorstep one night, shortly after his arrival in America, with a frightened-looking woman at his side. "He is carrying this large Iranian samovar which he wants to give me as thanks for bringing him to the United States," Morse recalled. "Then he proceeds to ask me if I've ever been with an Iranian woman before. He said, 'This is my sister—you can have her.'

"I said no thank you."

Bahrmand's story may be almost comical in some

respects, but Morse doesn't recall it lightly. "A lot of Iranians or Iraqis or Chinese who were sent to this country for training have gone back and are now working in institutions that are making biological weapons," he says. "I don't think that the person who was working for me is unique."

Dr. C. J. Peters of the CDC is an internationally respected infectious disease researcher and the author of the widely acclaimed book *Virus Hunters*. "Is it hard to grow these bugs?" he asks rhetorically. "Not if you're a microbiologist and so inclined. With smallpox, you can use an egg—you don't even need a cell culture. So far, the microbiologists have been nice guys, and so far, most terrorists know people who know how to make bombs, not grow viruses. But it's only a matter of time until that changes."

Of all of the nations that have developed biological weaponry, Russia remains the most troubling by far. Despite having signed the 1972 international Biological Weapons Convention along with 139 other nations, the Soviet Union ran the world's most aggressive research and development program for years. The flourishing Soviet program long remained hidden behind a cloak of secrecy and lies. Then Ken Alibek exposed the workings of the Soviet Biopreparat program in his chilling memoir, *Biohazard*.

Today Alibek has gained a measure of fame. Before writing his book, he was profiled in the *New Yorker* and testified about biological weapons before Congress. I first met Alibek before all of that—in December 1997, while author Mark Olshaker and I were researching an article about bioterrorism, following a piece on the topic I had published in *Newsweek*. It was a time when international attention was tightly focused on the hazards of the Iraqi biological weapons program and the risk of attack on the battlefield. But I was convinced, as I'll explain shortly, that the greater risk comes not from the nations that have developed such weapons, but from the smaller groups the technologies might leak to. A contact within the CIA told me that a man named Alibekov, the highest-ranking defector from the Soviet biological arms program, was living in the Washington area and had shortened his name to Alibek. A quick call to directory assistance and I had the number of a "K. Alibek" in northern Virginia. I called and left a message on his answering machine. Two days later a man with a heavy Russian accent returned my call and agreed to meet me in an Alexandria coffee shop.

As Olshaker and I sat in the coffee shop on that Saturday morning, we were expecting a severe Russian, the kind of cold robotic figure made famous in the Cold War James Bond films. We were surprised, then, when an unassuming man with dark, markedly Asian features stepped into the restaurant with his strikingly beautiful wife. Working across a language barrier, we talked for the better part of an hour, and I was horrified by his calm descriptions of the monstrous biological weapons he had created over the years: anthrax, smallpox, plague, Marburg, tularemia, and more. He talked about his concerns about the North Korean bioweapons efforts as well. It was the first of several discussions we would have. Later conversations with national security officials confirmed that everything he told us was true.

In our conversations and in his memoir, Alibek recounted heading up the job of "weaponizing" anthrax at Stepnogorsk after earlier successes with tularemia and brucellosis. In fact, Alibek pointed out with pride, he succeeded beyond the dreams of his superiors. Before being officially shut down in 1992, Alibek says, Biopreparat developed two thousand strains of anthrax alone, as well as bomb- and missile-ready versions of smallpox, the hemorrhagic fever Marburg, plague, and many more. Soviet scientists were even exploring the creation of genetically engineered strains of combination viruses that would defy conventional treatments, and proteins that would cause nerve damage or madness. They produced and replenished ready stockpiles of the most important agents, maintaining a 4,500-metric-ton supply of anthrax at all times.

That last figure stuns William Patrick, who served in the former U.S. biological weapons program in the 1960s. Bill Patrick says the U.S. stockpile of anthrax produced before the program was disbanded was 1 metric ton per year. "It's just hard for me to conceive of 4,500 metric tons of anthrax," Patrick says. "Can you imagine what that material could have done?"

The effectiveness of those Soviet pathogens was proved, tragically, at Sverdlovsk, the Russian city that had the dubious honor of hosting the nation's anthrax program until Alibek took over in Stepnogorsk. In 1979 a small amount of anthrax escaped from the facility's labs when apparently, a scientist neglected to replace a filter over an exhaust vent to the outside world. In all, seventy-seven cases of anthrax were identified with certainty; sixty-six patients died. The actual total number of cases was probably considerably more than one hundred. Although the cases occurred close to the facility, investigators later found anthrax deaths among sheep and cows in six different villages up to fifty kilometers away. If this plant had been built in a bigger town, many more cases would likely have occurred.

Although the Soviets long insisted that the Sverdlovsk epidemic was a case of tainted meat, and denied that the nation was still involved in biological weapons production, it was a far smaller mistake than it could have been: researchers who documented the outbreak calculate that the weight of spores released as an aerosol could have been as little as a few milligrams of powder. Russian President Boris Yeltsin finally admitted in 1993 that the outbreak was a military mistake.

Russia officially disbanded the biowarfare effort in 1992, but neither Patrick nor Alibek thinks that it was completely destroyed. Alibek testified before Congress that "considerable downsizing in this area did indeed occur," and included destruction of existing biological weapons stockpiles. However, there still remains doubt that Russia has completely dismantled the old program.

When the Soviet Union collapsed, Western policy makers and intelligence officials began to ask what would become of the vast biological warfare apparatus. Officially, the Soviets were dismantling the programs. But dismantling is not the same as destroying; facilities whose use has shifted from warlike means to peaceful ones can be turned back again. And as you'll see, the people who worked there don't suddenly develop amnesia.

Patrick, who debriefed Alibek after his defection to the United States, is of the same mind. "Knowing the paranoia" of the former Soviets, and now the Russians, he says, "I don't think they'd give it up. It's too effective a weapons system." Patrick notes that, after all, "the greatest speedup of their BW programs occurred *after* they signed the BW Convention! How can we believe these people when they tell us they've destroyed their seed stocks and their BW capability?" Every new day's news brings fresh uncertainty and worries; with Yeltsin's resignation on the last day of 1999 and the ascension of the mysterious Vladimir Putin, we entered yet another phase in the evolution of the former Soviet Union—and no one knows what to expect.

You might think that the implosion of communism and the promise from Boris Yeltsin to stop producing bioweapons might have lessened the menace. In some ways it did, since the decline of the Soviet Union cut off some of the support and protection for left-wing terrorist groups. But terrorism did not go away, and neither did state sponsorship of it.

Iraq, too, has run a large-scale biological weapons program. At the height of its preparedness, in 1991, Iraq had twenty-five missile warheads filled with anthrax, botulinum toxin, and the cancer-causing agent aflatoxin, as well as some three hundred biological bombs, according to congressional testimony by Ambassador Rolf Ekeus, who headed the United Nations special commission that investigated Iraq's arsenal after the Gulf War. The CIA determined that the Iraqi biological warfare program produced 85,000 liters of anthrax, 380,000 liters of botulinum toxin, and 2,200 liters of cancer-causing aflatoxin, among other agents, and developed missile warheads, aerial bombs, and aircraft-mounted aerosol spray tanks. And while Saddam

Hussein was willing to give up his ability to manufacture nuclear weapons, his cat-and-mouse game with U.N. inspectors shows how fiercely he has tried to hold on to the ability to manufacture and use biological agents. To date, the United Nations has not identified or destroyed any of Saddam Hussein's biological weapons, and they are likely still out there, waiting to be used—either as weapons of war by Iraq or, worse, as weapons of terror by unknown third parties to whom the agents or technologies could all too plausibly leak.

China has come under special scrutiny from specialists in weapons of mass destruction because it has proved willing—even eager—to sell destructive technologies it develops to others. The Deutch commission classified the prospect of the "transfer of nuclear, chemical, and biological weapons, delivery means, and technology by China" as one of "the most serious threats" to the United States, its military forces, and its vital interests abroad.

The effect of using the agents in a stealth attack against civilians would be far worse than on the battlefield. Even if a conflict did arise in which the weapons might be used, nations at war are at least partially able to prepare their troops for attack through vaccination and distribution of protective and medical supplies. The difference between those two situations can be illustrated with an analogy to sports. When professional football players sack the opposing quarterback, he is pumped up, trained in the hard business of collision, and equipped with a full set of pads— and unlikely to be badly injured. Now imagine the same scenario, except the players are suddenly crashing into an unsuspecting Woody Allen as he's playing his clarinet at Michael's Pub.

Other state bioweapons programs bring up issues of their own. Take North Korea, which has by many accounts been working up a biological weapons effort. The nightmare scenario in North Korea would involve an accidental outbreak from a bioweapons lab similar to the accidental outbreak of anthrax in the Soviet Union's Sverdlovsk plant. If a similar accident led to the reemergence of smallpox in an isolated, impoverished nation like North Korea, the highly contagious disease could quickly spread out of control. In January 1999, while attending a meeting at the White House, I asked a high-level administration security official whether the United States would send its precious (and limited) supplies of smallpox vaccine to end a North Korean disaster. The answer was quick: "Absolutely not." But the price of not acting early in an emerging smallpox outbreak will precipitate a global calamity—and the work of a generation of smallpox fighters to eradicate the ancient plague could be undone.

The Middle East, as always, presents a uniquely complex set of problems. I had the opportunity to understand the gravity of this issue at the international level in the fall of 1998, when I was asked to advise His Majesty King Hussein of Jordan on biological terrorism issues. The king's physician at the Mayo Clinic, Dr. Walt Wilson, is a longtime friend and respected colleague. As a result of Walt's subsequent conversations with His Majesty and a lecture I presented to his son, the prince, and others of his senior staff, I was summoned to meet with the king at his estate outside of London. One of the world's greatest experts on terrorism—he lived with the threat of it, after all, on a daily basis—he deeply understood the likelihood of such an event, and the possible consequences.

The ailing monarch was bald from chemotherapy and weakened by disease, but the quiet force of his personality was undiminished. Queen Noor and the head of Jordanian intelligence also sat in on the discussion. During our conversations, the Hashemite king explained to me that his country was surrounded by nations that he believed were developing biological weapons—Syria, Iraq, and Israel. Any disease outbreak—whether caused intentionally, by terrorist event, or by accident—would be chalked up to government action in such an overheated political climate. Anger, recrimination, and a return of warfare to that troubled part of the world were almost sure to follow.

Just days before his sudden and untimely death, the king wrote a letter to his brother that laid out his fears. "After a thorough examination, it is clear to me that the situation has become extremely dangerous and is a source of constant concern to the world," he wrote. "Perhaps biological weapons are the most dangerous of all, because they reintroduce to the world certain diseases that have already become extinct, like smallpox. . . . In the present time of fast communication, such disease can move with an amazing speed around the world, ending the lives of all people without discrimination." He was prepared to raise the issue to the top of the international agenda. His death resulted in the loss of a voice of leadership we desperately needed.

OTHER TERRORIST GROUPS

Violent groups fueled by a broad variety of ideologies have presented new challenges to those who monitor terrorist acts. Religious extremists such as the network affiliated with Saudi financier Osama bin Laden have proved willing to commit large-scale destructive acts, including the U.S. embassy bombings in Tanzania and Kenya. Further, millennialist cult groups, which espouse the view that the world will end during or soon after the year 2000, grew in prominence as the calendar neared the changeover. Some even believe it is their duty to help bring about the apocalypse—like Concerned Christians, whose members disappeared from their home base in Denver in 1998. Fourteen members of the cult were later expelled from Jerusalem, in January 1999, on suspicions by authorities there that the group was planning to provoke a bloody gunfight that members believed would bring about the reincarnation of Christ as their group's leader, Monte Kim Miller.

In October 1999 the FBI issued a report, "Project Megiddo," to warn law enforcement around the nation that domestic terrorism tied to the coming of the year 2000 was increasingly likely. "Militias, adherents of racist belief systems such as Christian Identity and Odinism, and other radical domestic extremists are clearly focusing on the millennium as a time of action," the Bureau said. "Armed with the urgency of the millennium as a motivating factor, new clandestine groups may conceivably form to engage in violence toward the U.S. government or its citizens." The report identified a long list of potential targets, including military facilities, United Nations buildings and workers, and organizations for African Americans, Jews, homosexuals, and other minorities, and suggested that the risk of terrorism could extend into the year 2000 and beyond.

A number of extremist groups hold the view that this corrupt planet is due for what Jessica Stern calls a "cleansing apocalypse." She notes that Aum Shinrikyo leader Shoko Asahara taught his doomsday cultists that "in the coming conflict between good and evil they would have to fight with every available weapon." (Similar views help explain the appeal of survivalism for the white-supremacist group known as Christian Identity, Stern noted.) Such groups may believe that in order to help bring about the end of days, they need a suitably effective weapon that will prepare large populations over big geographic areas to meet their makers. Bombs, chemical weapons, and bullets may not be enough to accomplish that goal, while the utilization of plagues has a biblical overtone that might resonate with them.

NEW WORLD TERRORISTS

In no small part, the United States has shaped the tactics of its enemies through its very strength. America's role as the world's sole remaining military superpower has dissuaded any potential enemy from taking the nation on frontally. "American military superiority on the conventional battlefield pushes its adversaries toward unconventional alternatives," Deutch and his coauthors wrote in the *Foreign Affairs* "catastrophic terrorism" article. The biggest missiles and the strongest armor offer no protection from microbes.

A weapon of mass destruction is fundamentally more dangerous in the hands of a terrorist group than in the hands of even a rogue nation. A nation may produce such weapons as a threat, a bargaining chip, or simply for the sense that the strategic balance can be equalized in case of confrontation with a more powerful foe. A terrorist group has no need for such nuances and contingencies: the groups make their statements not by *threatening*, but by *doing*.

Two groups exemplify the new wave of terrorist groups: the Japanese cult Aum Shinrikyo and the constellation of terrorists around Osama bin Laden.

Aum Shinrikyo, founded in the 1980s by Japanese businessman and former yoga instructor Shoko Asahara, presents a clear warning of the kind of organization that can be assembled and the tools they might have at their command. A U.S. Senate investigation concluded that at the time the group mounted its Tokyo subway attack in 1995, it had some fifty thousand members worldwide and controlled assets of $1.4 billion; its ranks included scientists and engineers from Japan, Russia, and elsewhere. Motivated by a belief that the coming millennium would bring nuclear war, Asahara began ordering his believers to purchase, develop, and deploy every kind of weapon of mass destruction, including nuclear, chemical, and biological weapons.

The group was apparently unsuccessful in pulling together nuclear weaponry, but its record in accumulating chemical and biological weapons was far better, evinced in the nine biological attacks before the Tokyo subway incident. The group tried to spray botulinum toxin and *Bacillus anthracis* near the Japanese Diet, the Imperial Palace, and at the American naval bases of Yokosuka and Yokohama, among other sites. All of the biological attacks were failures; no humans lost their lives. Seiichi Endo, the biological arms chief for the cult, has since said that the anthrax's weakness was due to the fact that he had mistakenly acquired a vaccine strain that was not virulent at all. In one attack, a group member reportedly had pangs of conscience and neglected to arm the weapons.

Only when they attempted to use chemical weapons did they begin to see success, as in a June 1994 attack at the town of Matsumoto, now credited to the group, that left seven people dead and two hundred ill. Even the March 20, 1995, attack in Tokyo was a rush job, since Japanese law enforcement authorities were finally catching up with the group's activities. The Aum did not rely on fancy dissemination equipment for the highly volatile gas; Aum Shinrikyo members simply filled eleven plastic containers with sarin, placed them on the floor of subway cars, and punctured the bags with umbrella tips, letting Parkinson's law do the rest.

"Aum Shinrikyo brought everything out in the open," says Gordon Oehler, former director of the CIA's Nonproliferation Center. "The unthinkable became thinkable. Anybody can bring the battlefield to our porch now."

Some experts draw comfort from the relative incompetence of the Aum Shinrikyo attacks, saying that it shows the difficulty of mounting effective attacks, even with vast resources, expertise, and manpower. Others have not been so sanguine, however. "If Aum had taken more time and been more proficient it might have killed thousands or even tens of thousands," wrote Stefan Leader, a terrorism specialist with Eagle Research Group Inc. of Arlington, Virginia, in a 1999 essay in *Jane's Intelligence Review*. Despite the twelve deaths and five thousand injured in the attack, Stern said, the cult intended to kill thousands more, but simply ran out of time: "The 1995 poison-gas attack in the Tokyo subway was carried out in haste and did not represent the cult's full potential."

At the time of the arrests, the group had stored away enough VX to kill about fifteen thousand people, and hundreds of tons of chemicals used to make sarin. Despite the group's initial lack of success with biological agents, it had stored a "large amount" of *Clostridium botulinum* and 160 barrels of growth media. What's more, members of the group had reportedly traveled to Zaire to collect samples of the Ebola virus; it was also working with strains of Q fever, which appear to have infected some members of the cult. In 1997 then CIA Director Deutch testified before U.S. Senate hearings on the proliferation of weapons of mass destruction that the group planned attacks against the United States. Fortunately, the group ran out of time.

When discussing the failures of Aum Shinrikyo, it's important to keep in mind that the past doesn't dictate the future—or, as the financial prospectuses for mutual funds say, past performance is no guarantee of future results. The underlying pool of knowledge is always advancing. Making broad pronouncements about the future of biological terrorism based on what's happened so far is as limited as comparing the ability of soldiers to kill with World War I–era technology to the abilities to kill with today's weapons.

Though the sect was supposedly liquidated by government order, there are signs that it is back, "regrouping, recruiting new members at home and abroad, and raising vast sums of money," according to security officials and Japanese terrorism experts quoted in *The New York Times*. In other words, they're not gone at all. And while the group in early 2000 disavowed Asahara and changed its name to "Aleph," skeptics claim the disavowal was a ruse to avoid penalties under new Japanese law as Asahara's prison term winds to a close. They warn that the group has not changed its ideals at all and could pose as great a threat as ever.

Although the centralized groups are fading, there is no shortage of well-funded multinational terrorist organizations that have risen up to take their places. The best known of the current crop of terrorist organizations, and the best example of the shift from centralized to decentralized terror, is the organization headed by exiled Saudi millionaire Osama bin Laden, which has been linked to the bombing attacks on embassies in Dar es Salaam, Tanzania, and Nairobi, Kenya, which killed 224 people and injured thousands.

In the past, Osama bin Laden has said that he is targeting the American military, not its citizens. But his line has toughened—and his scope broadened—since those bombings. In a January 1999 interview, he announced, "If the majority of the American people support their dissolute president, this means the American people are fighting us and we have a right to target them. Any American who pays taxes to his government is our target because he is helping the American war machine against the Muslim nation." He loudly criticized Israel and the United States for policies that he says have caused the deaths of innocent civilians, and stated that the deaths of innocent civilians in the jihad bombings—even Muslims—were "permissible under Islam," since it is impossible to "repel these Americans without assaulting them." As for whether he was acquiring chemical and biological weapons, he would only say, "Acquiring weapons for the defense of Muslims is a religious duty. If I have indeed acquired these weapons, then I thank God for enabling me to do so. And if I seek to acquire these weapons, I am carrying out a duty. It would be a sin for Muslims not to try to possess the weapons that would prevent the infidels from inflicting harm on Muslims." Fighting, he explained, "is a part of our religion and our Shari'a. . . . Hostility toward America is a religious duty, and we hope to be rewarded for it by God."

Terrorism specialist Leader writes that bin Laden's comments "reveal a man who believes he is on a mission from God to evict the USA from the Middle East and is prepared to kill not only military and government personnel but innocent civilians as well."

Bin Laden also shows how American authorities sometimes seem not to grasp the changes that have taken place in terrorist groups. According to *Washington Post* reporters Colum Lynch and Vernon Loeb, the American indictments against bin Laden and sixteen codefendants describe bin Laden "as the leader, or 'emir,' of al Qaeda, a 'global terrorist organization' with tentacles that allegedly reach from his hideout in the mountains of Afghanistan to followers in Texas, Florida and New York." But in fact, the reporters write, "the portrait of al Qaeda that emerges from hundreds of pages of court filings looks less like a tight-knit group under one man's command than a disjointed, shadowy confederation of extremists from all over the Islamic world." Terrorism experts and government officials round out the picture, the reporters write: "It remains unclear whether al Qaeda has managed to assemble a powerful and dangerous network inside the United States or merely a sprinkling of sympathizers whose links to one another, and to bin Laden, are often tenuous." In fact, Brian Jenkins of Kroll Associates told them, "to imply something too organized, too hierarchical, misses the reality. . . . We're obliged to talk about universes of like-minded fanatics — nothing that appears like a wiring diagram Western bureaucrats are familiar with."

The Egyptian militant group Islamic Jihad, affiliated with bin Laden, has devised both chemical and biological weapons. The group has planned out "100 attacks against US and Israeli targets and public figures in different parts of the world," according to a statement by Ahmed Salama Mabruk, former head of the jihad's military operations. The newspaper *Al Hayat* interviewed Mabruk just before he was sentenced in a mass trial for jihad members. According to the Agence France-Presse account of the story, Mabruk said the CIA in Azerbaijan took a computer disk with the plan from him upon his arrest. Mabruk "said he

assumed that, upon learning of his arrest, [bin Laden] and
his lieutenant Ayman al-Zawahri would have changed the
dates and locations for the attacks outlined in the plan."

LONE WOLVES

Perhaps the most troubling development in the changes
that terrorism has undergone over the years is the rise
of the "lone wolf," people like "Ed" in the scenario that
opens this chapter. Some are self-taught, some taught by
institutions and the government. Many are motivated by
racial and religious hatred and commit unspeakable crimes
with no apparent central authority, even more decentralized than groups like bin Laden's. Compared to members
of more structured terrorist organizations, the lone wolves
are nearly impossible to track.

The modern terrorist can come from anywhere, picking
up influences on the fly and molding them into a philosophy. Some experts refer to the new breed as the cowboy
terrorist. Increasingly, they arise seemingly from nowhere,
responding not to direct orders, but to a kind of ad hoc
collection of influences and opinions. Unabomber Theodore
Kaczynski read widely from American and European phi-
losophers and historians, knitting their ideas, together
with environmentalist and political tracts, into his manifesto. That document includes discourses on technology,
politics, psychology, sociology, spanking, and the perceived
evils of the Sylvan Learning Centers, which provide tu-
toring for schoolchildren. Kaczynski saw himself as a terrorist trying to start a revolution to stop the spread of
technology.

Whatever the nature of the groups, it is becoming increasingly clear that a cataclysmic event can emerge from a loosely knit organization—or from no organization at all. Timothy McVeigh did not appear to be affiliated with any larger group, though his interest in antigovernment groups was clear in his choice of reading material. He was a fan of *The Turner Diaries*, a racist novel that describes a bombing attack on FBI headquarters and could have served as a blueprint for McVeigh's own actions, and of *Hunter*, novels written by William Pierce, head of the Neo-Nazi National Alliance. In her closing argument in the case, U.S. Attorney Beth Wilkinson said *The Turner Diaries* "was the same book McVeigh carried with him everywhere he went. In fact, you heard that he even carried it to field maneuvers way back in 1990."

A passage from *The Turner Diaries* highlighted by McVeigh comments on the reasoning behind bombing the U.S. Capitol and an Israel-bound jetliner: "The real value of our attacks today lies in the psychological impact, not in the immediate casualties. For one thing, our efforts against the System gained immeasurably in credibility. More important, though, is what we taught the politicians and the bureaucrats. They learned today that not one of them is beyond our reach. They can huddle behind barbed wire and tanks in the city, or they can hide behind the concrete walls and alarm systems of their country estates, but we can still find them and kill them."

This new generation of homegrown terrorists is auditioning for parts in a drama written in their own minds. Take another recent "lone wolf," Buford O. Furrow, who epitomizes the breed. Furrow is the white supremacist who opened fire on adults and children in August 1999 at a Los Angeles Jewish community center and then killed a postal worker before his arrest. Furrow "may represent the wave of the future: a new generation of racial warriors who believe in acting alone," writes Hanna Rosin of the *Washington Post*. Furrow appears to have built a philosophy from the rantings of the Aryan Nations, a group that follows the theology known as Christian Identity, and a movement called the Phineas Priesthood. Christian Identity teaches that racial minorities were created before Adam and Eve and are little better than animals; Jews, they say, are the race spawned by the coupling of Eve and Satan. Members of the Phineas Priesthood hold similar white-supremacist views, with a deadly twist: members allegedly attempt to prove themselves through acts of terrorism.

Lynchburg, Virginia, author Richard Kelly Hoskins coined the "priesthood" in his 1990 book, *Vigilantes of Christendom*. Hoskins named the movement for the biblical figure who kills one of his tribesmen for marrying out of his faith. Hoskins laid out the concepts, however, in an earlier book, *War Cycles/Peace Cycles*, a copy of which was found in the van that Furrow abandoned in Los Angeles. Furrow, like Kaczynski, has been diagnosed as seriously mentally ill, but mental illness does not reduce people's capacity to do evil things, and may even remove the moral strictures that prevent the rest of society from performing such acts. Indeed, "normal" people do not do these things. But the people who do them need not have been diagnosed with a mental illness. Mike Reynolds of the Southern Poverty Law Center says that although some people who are mentally ill join these groups, there are many,

many others who are "clear-headed" people whose belief systems allow them to kill. "You can't say everybody in the mujahideen, everybody in Islamic Jihad, everybody in the Hezbollah who have committed acts of terrorism is mentally incapacitated." Anything that makes human beings "the other" or "the enemy" helps them to commit their crimes, he explains; in fact, such belief systems have been used to teach people how to kill for centuries: "That's how we create soldiers."

Some experts believe that this process of dehumanizing others has become an inadvertent undercurrent of the media influences bombarding youngsters, including violent films and video games. David Grossman, who teaches psychology at Arkansas State University and directs what he calls the Killology Research Group, studies what allows people to get over the natural and ingrained moral objections to killing. A retired army officer, Grossman refers to historical statistics to show that even in World War II, soldiers fired their weapons only 15 percent of the time that they had a clear shot at the enemy. The American military spent the ensuing decades developing methods to condition soldiers to overcome their reticence—such as changing target practice from firing at abstract targets to firing at increasingly realistic ones. According to Grossman's research, that and other methods of conditioning helped to raise the "kill rates" among American soldiers to 90 percent by the time of the Vietnam War. Grossman argues that today's violent video games are effectively conditioning young people to get over their objections to killing and make the act seem commonplace, even fun. The view has drawn argument and ridicule from many quarters, especially civil libertarians who see this view as an excuse to weaken First Amendment protections of speech. But Grossman—who lives in Jonesboro, Arkansas, site of the 1998 student shootings that killed five people and wounded ten, and a hired expert witness for the parents of the three girls killed in the 1998 West Paducah, Kentucky, school shooting—is adamant in his belief.

People who already have a tendency to commit violent acts are drawn to ideologies of hate, Reynolds says. "When you have belief systems, whether religious or political, that advocate violence as the means necessary, you're going to attract individuals to that group," he says. "Once they're given a belief system and a support system that justify their behavior, there's no holds barred. That's like giving them a green light."

Others associated with the Phineas Priesthood include four Idaho men who set off bombs in July 1996 at a Spokane, Washington, Planned Parenthood center and signed a letter with the symbol of the Phineas Priesthood; in 1991 prosecutors linked the killer of civil rights leader Medgar Evers, Byron de la Beckwith, to the group as well. Other accused cowboy terrorists are Benjamin Matthew Williams and his brother James Tyler Williams, arrested in connection with the shooting of a gay couple near Redding, California, in July 1999; the two are also suspects in fires that did nearly $1 million in damage to three Sacramento synagogues in June 1999. And then there is the sad case of Benjamin Nathaniel Smith, whose shooting spree against blacks, Asians, and Jews over the Fourth of July weekend in 1999 killed two people and injured nine before he committed suicide.

Would such people turn to biological weapons? They

already have. Experts who monitor American hate groups such as Aryan Nations say that members of those groups have discussed using biological weapons and toxins since the 1980s.

Just as bioterrorism has changed the nature of terrorism, it has also changed the nature of the people who engage in it. These are people with special skills, having taken college-level courses in microbiology or other disciplines helpful in tackling the technological challenges of biological terrorism. Many Aum Shinrikyo cultists had earned advanced science degrees and worked in the fields of science and technology.

Then there is Larry Wayne Harris, a trained Ohio microbiologist and member of Aryan Nations, who was arrested after having obtained by mail three vials of the bacteria that cause bubonic plague. He later told *U.S. News & World Report* that his comrades will use biological weapons against government officials and whole cities if provoked, and intend to use such weapons if they will help them reach their goals of a separate nation. "If they arrest a bunch of our guys, they get a test tube in the mail," he said. "How many cities are you willing to lose before you back off?" he asked. "At what point do you say: 'If these guys want to go off to the Northwest and have five states declared to be their own free and independent country, let them do it'?"

Harris reappeared in February 1998 when he and William Joe Leavitt were arrested in Las Vegas on suspicion that they possessed anthrax. The FBI said Harris bragged in Las Vegas that he carried enough anthrax to "wipe out the city" and had previously planned to conduct an Aum Shinrikyo–type attack on the New York City subways. In fact, authorities later determined that the two were carrying a relatively harmless form of the bacteria in a veterinary vaccine. Harris said he was working on anti-anthrax medicines; others said he may have been trying to engineer it into a lethal strain.

A 1984 outbreak of salmonella in The Dalles, Oregon, that sickened 750 people was later traced back to members of the Rajneeshee cult. Members of the cult, hoping to influence the outcome of a county election, had spread the bacteria through ten local salad bars.

Then there is the case of Thomas Lavy, who was stopped in 1993 at the Alaska-Canada border by Canadian agents. The border officials found four guns, twenty thousand rounds of ammunition, neo-Nazi literature, how-to manuals for making biological and chemical weapons, and a plastic bag filled with ricin. "Who was he taking that stuff to when he came out of Alaska, anyway?" asks Mike Reynolds. "He said he was going to kill coyotes on his chicken farm—yeah, right." Lavy has since hanged himself.

Reynolds says one of his biggest worries is the possibility that America's white-supremacist groups will develop strong bonds with similar racist organizations in Russia. "We have a lot of work to do," he says.

3. WHERE THEY WILL ATTACK

Down in the stadium, no one even sees the airplane passing to the north; they are focused on the game below, not the quiet disaster above. The roar of the crowd drowns out the growl of the plane's engine. By the time the disperse cloud of dust reaches the ground, it's imperceptible; at two to six microns in size, the spores are smaller than pollen—as many as twenty thousand of them, lined up, would barely stretch an inch. There is nothing to see, and the spores smell of nothing. The spectators, athletes, the billionaire owner in a brief foray out of his skybox—all take the tiny killers deep into their lungs.

Once a spore hits the lungs, the victim's immune system

sounds an alert. Pulmonary macrophages—the cells that eat many of the body's intruders—swarm to the site, and one envelops the spore. That's the last word on most invasions; macrophages generally eat whatever bacteria make their way into macrophage territory. But the macrophage can't digest the spore's rugged coating; instead of killing the unwelcome guest, the macrophage becomes its chauffeur. The macrophage carries the spore through the body's network of blood and lymph vessels, finally nestling in the lymph nodes of the mediastinum, the area just behind the breastbone.

As the spore reaches the warm, wet tissues of the lymph nodes, it knows that it has found a new home. In some of the victims, a signal goes out right away to execute the instructions stored in its genetic code. In others, the spores bide their time, waiting days or even weeks before starting their programs.

As the age-old DNA program boots up, it begins a process of germination that transforms the dry spore into a living, growing bacterium: eating and multiplying, eating and multiplying. All the while, another portion of the cell's internal software sets up a microscopic factory that begins pumping out the toxins—a handful of chemicals that will snip the MAPKK protein and wreak havoc throughout the body.

Of the 74,000 people attending the game that night, 39,000 are infected. In the surrounding neighborhoods, 15,000 more are infected after inhaling the drifting spores. After the game, the fans go back to their homes. Because it is late summer, vacationing families have traveled from all over the country to attend. Over the next day or two, they leave their hotels and return to their homes, scattering across the nation and the world.

The bacterial cells continue to reproduce. After a day or two, many of the victims begin developing a fever and dry cough;

some complain of shortness of breath and chest pain. Most of them pop a few Tylenol or Advil and try to ride out the illness. Some see their doctors and are told to take it easy and drink plenty of fluids. Some of the toughest cases show up at emergency rooms. Many patients begin to feel better after a couple of days; only later will they slip back into serious illness.

Other than the unusually large number of patients, there's nothing special about the disease at this stage to raise suspicions. But the victims are seeking help from so many different health care providers that the true scope of the epidemic initially goes unnoticed. "Looks like you've caught a bug," the ER doctor tells a middle-aged man. "Lot of that going around. It's probably what we call an enterovirus." He orders up an X ray just to be on the safe side, but the film shows no signs of pneumonia. He puts it aside, missing a telltale sign of an anthrax infection—a widening of the mediastinum caused by the burgeoning bacterial population.

Within a week of the attack, more than 20,000 people have swamped local hospitals, and the medical establishment and local government are trying to figure out what has happened. The city, even the region, is in utter chaos. Panic has overrun both the government and the medical community. The most susceptible victims rapidly spiral into the agonizing crash of respiratory shutdown, with high fever and shortness of breath; many turn blue. Hundreds—only the first hundreds—die. Desperate calls go out for help to the state health department and to the Centers for Disease Control and Prevention in Atlanta. There's little time to discuss the epidemic, however, because the number of cases quickly overtakes the ability of local hospitals to cope.

On the seventh day after the attack, a lab is able to culture anthrax from the blood of a young victim. Since antibiotics cannot help patients once symptoms begin to appear, physicians are told to hold their supplies for those who are most in need and likeliest to survive—and even then, it's not enough.

When the mayor of the city goes on television to announce that anthrax has been loosed upon the town, the level of fear skyrockets even higher than before. It's the biggest news story on the planet, with round-the-clock coverage. Inaccurate news reports and rumors lead the citizenry to believe that the disease is highly contagious, and the city shuts down as its people shut in.

Across the country, television viewers tune in to watch live reports delivered by reporters in gas masks. They are shown mob scenes at hospitals and fistfights at pharmacies rumored to have antibiotics on hand. After the pharmaceutical companies report that a new supply would take weeks to manufacture, a black market explodes, with drug dealers charging more for antibiotics than for heroin or cocaine. Health authorities take over the city's convention center, turning it into a gigantic anthrax ward. But the need for thousands of beds quickly overwhelms that facility as well.

As the tenth day after the attack dawns, there are 30,000 sick and 700 dead; by the next day those numbers will have jumped to 38,000 sick and 1,500 dead. Doctors' offices, clinics, and emergency rooms continue to be overwhelmed with people demanding antibiotic treatment.

Once the mysterious outbreak is identified as anthrax, the Centers for Disease Control and Prevention begins to send antibiotics from the national pharmaceutical stockpile. Within 36 hours of activating the stockpile, and making some quick turnaround plans with drug manufacturers for additional medicines, there are enough doses of antibiotics on hand to treat, in a perfect world, more than 5 million people.

It's a help, but the world is not perfect; getting the drugs to people turns out to be a logistical nightmare. With assistance from other city and state officials, overwhelmed public health authorities open storefronts to distribute antibiotics, but there aren't enough warm bodies to help distribute them efficiently. Many of the people who might be expected to help with this operation are too sick with anthrax even to help themselves, let alone others. Lines at the distribution centers stretch around the block, with angry, frightened people pushing and shoving to get the lifesaving drugs. Lines give way to mass unorganized crowds; soon they give way to frightened mob scenes. Several distribution centers have to be shut down after fights turn into riots.

In cities nationwide, anyone who comes down with a headache or chills decides he has anthrax. Local and state authorities around the country, not sure if they too are under attack, begin demanding that the stockpile be spread around more broadly; the 5 million treatments are now grossly inadequate.

The antibiotics are most effective if given before symptoms appear and are of little use to those who have full-blown illness. The drugs are also not as effective when given alone as they would be if given with the anthrax vaccine, but there is no vaccine for civilians. Still, the number of anthrax cases does not rise nearly as quickly, or as high, as it would without the antibiotic stockpile.

When the outbreak is finally linked by health officials to the football game and its surrounding areas, and the frightened populace begins to accept the idea that anthrax is not contagious, the hysteria becomes more focused, if not actually contained. Still, the illness relentlessly extends its reach. Just when it seems the numbers may never stop rising, about the end of the third week the population of those infected appears to flatten off at about 50,000 cases. The death toll shows no signs of slowing, however, as more of the sick go into their final decline. By now 18,000 people—a town's worth—have died. Eventually, 20,000 will lose their lives; it is the worst man-made disaster in the history of the world.

As the crisis grows, police and intelligence officials look for the culprit. Terrorism experts raise every possibility, from Islamic fundamentalists to "lone wolf" killers, but no one has a clue. Raids on suspected terrorist organizations yield dozens of arrests, but no anthrax is found. A local Iraqi refugee family is held in the county jail for six days without explanation; the news media learn of the arrest, and speculation rages that Saddam Hussein is responsible for the attack.

But Ed, who has never been affiliated with a terrorist organization, never run afoul of the law, never even held forth at a local watering hole about his plan or his views, remains an invisible man. He sits at home, in front of the television, admiring his handiwork.

THIS GRIM SCENARIO and the mind-boggling estimates of death and illness caused by the airborne attack might seem outrageous—how, after all, could a single act wipe out as many people as are found in whole towns? But it is, in fact, a middle-of-the-road view of the kind of damage that an anthrax attack might cause. Remember that the Office of Technology Assessment estimated that just a couple of hundred pounds of anthrax spores released on a clear, calm night upwind of Washington, D.C., could kill 1 to 3 *million* people. This scenario is precisely the kind of case study that policy makers and public health officials are now beginning to give careful consideration.

In fact, many of the elements of the preceding scenario were drawn from a 1999 presentation at a bioterrorism conference sponsored by the Johns Hopkins Center for Civilian Biodefense Studies and the Infectious Diseases Society of America. I served on the planning committee for the conference, and we wanted to lay out in black and white what America could expect if an anthrax release occurred. Dr. Thomas Inglesby of the Johns Hopkins Center developed and delivered the scenario to the stunned audience. Inglesby's estimates of damage and death were slightly lower than those in the scenario you just read, but Inglesby later admitted to me that his numbers could have been too low—and that nobody yet knows how high they could go. He noted dryly at his talk that "such an epidemic would create extraordinary challenges for a modern American city."

It would not be the first time that biological weapons were used against a civilian population. Warfare with biological weapons has been employed many times, dating back to our earliest history. Hannibal, in preparation for a naval battle against King Eumenes of Pergamum in 184 B.C., ordered that earthen pots filled with "serpents of every kind" be hurled onto the decks of the Pergamum ships. During the Middle Ages, victims of infectious diseases became actual weapons. In 1346 the Tatar army attacked Caffa (in modern Ukraine), a highly fortified seaport, and catapulted bodies of dead plague victims over the city wall. An epidemic of plague followed. Genoese mercenary forces retreating from the city continued to spread plague throughout Europe.

Smallpox was used as a biological weapon in the New World among a population that had previously been free of the disease. In 1763, smallpox-laden blankets were used by the English at Fort Pitt and given to the Indians loyal to the French. The impact of contaminated blankets is unclear, however it is likely that the "raging epidemic" of smallpox among the Indians at that time was related.

By the twentieth century, biological warfare had become a science. The Germans were alleged to have used a variety of animal and human pathogens in World War I. Recent proof of that was found in police archives in Norway. Nineteen sugar cubes, each with embedded powder-filled glass vials, were discovered. They had been taken from the luggage of Baron Otto Karl von Rosen when he was arrested for espionage and sabotage. The glass vials had been stored for more than eighty years among other war-related artifacts. It is believed that the German aristocrat had intended to place the vials in hay in areas where commercial traffic across northern Norway used horses and reindeer to deliver British supplies to the front lines in the war with Germany. It was anticipated that when the vials were chewed and broken open by the animals, cases of gastrointestinal anthrax would result. This is just one example of numerous German biological sabotage programs during the war.

During World War II, the Japanese were accused of using biological agents against the Soviet Union, Mongolia, and China. In 1940 an outbreak of bubonic plague occurred in Cheking province, China, after a Japanese plane dropped ceramic pots loaded with contaminated rice and fleas; plague had never previously been recorded in the area. Similar outbreaks were reported to have occurred in eleven other Chinese cities following similar attacks.

Japan's Imperial Unit 371 used at least three thousand prisoners of war as guinea pigs. It is estimated that more than one thousand of these prisoners died in experiments with pathogens causing anthrax, botulism, brucellosis, cholera, dysentery, meningococcal infection, and plague.

Following World War II, biological weapons programs in the United States, Canada, the Soviet Union, and the United Kingdom continued to expand until 1972, when the Biological and Toxin Weapons Convention (BWC) was signed and ratified by 140 nations. Most notably, nations signing the BWC included the Soviet Union and Iraq. The agreement called for termination of all offensive weapons research and development as well as the destruction of existing stockpiles of agents. The United States and other Western nations complied, but today we know that at least ten nations did not—and that some actually greatly accelerated their research and development efforts after signing.

While the prospect of biological warfare is still a concern, it has been surpassed by the fear of biological terrorism against unsuspecting civilians. A bioterrorist has no end of rich, vulnerable targets in today's world; one could even say that our society has evolved in ways that enhance the probable success of a biological attack. Any place can be a target, although some common characteristics make some places more likely than others.

Most important, we are seemingly always part of a crowd. "Modern societies are particularly susceptible to weapons that are capable of killing many people at one time. . . . Their citizens tend to live, work, and travel in close proximity, providing concentrated targets," expert Stern warns in *The Ultimate Terrorists*.

Some crowds appear to be better than others in a terrorist's eyes; the higher the profile of the attack site, the more the attraction, the FBI stated in its 1996 report on terrorism. Transportation centers are a perennial security concern, wrote the FBI, because "their very nature and design" present an attractive target to terrorists, "where the objective for inflicting mass casualties can be obtained." With biological terrorism, an additional factor comes into play: the fact that the victims are on the move, dispersing around the country or even around the world. The effects of a biological attack will be amplified by the spread around the world; an airport attack using a highly contagious agent such as smallpox will create multiple outbreaks worldwide.

The value of a target to terrorists is also apparently increased if it has symbolic value, which raises the visibility of the act. The World Trade Center, of course, is among the most famous office complexes in the world's most famous city. Similarly, the bombing attack at the XXVI Olympiad in 1996 in Atlanta occurred at a symbol of international cooperation and in the presence of record-breaking crowds. The FBI's 1996 terrorism report suggested that the massive celebration of global athleticism and cooperation provided the apparently irresistible combination of huge crowds and high symbolism, "a powerful motivating force for individual zealots or terrorist extremists to use these events as staging areas for their causes."

Not all symbolic targets are obvious. Timothy McVeigh set off his truck bomb in front of a completely unremarkable federal office building in Oklahoma City—reasoning in part that security would be lower at a facility that was not in one of the nation's great urban centers.

Ultimately, the biggest crowds are to be found in our cities and the areas surrounding them. Unlike our rural forebears, we have flocked from the countryside to large metropolitan areas, where our great numbers increase the efficiency of a potential attack. Increasingly, America is an urban nation. In 1900 there were only 76,212,168 people in the United States; 39.6 percent of the nation lived in urban areas and 60.4 percent lived in rural areas. During the 1920s, that ratio evened out and then began to shift to a preponderance of urban Americans, up to 75 percent by 1990.

Not everyone lives in the center of these towns, of course; the tightly packed urban core has been largely replaced by sprawling metropolitan regions with pockets of built-up "edge cities" sprouting at the periphery. The journalist who first applied the term to these outer metropolises, Joel Garreau of the *Washington Post*, calls the appearance of the edge city "the biggest change in a hundred years in how we build cities." Vast, spread out, and prosperous, these highwayside clusters of office towers and megamalls and suburbs, each feeding on the other, can be found along Route 128 in Boston and at Tyson's Corner, Virginia, at the rim of the Washington Beltway, at Houston's Galleria, and at Atlanta's Perimeter Center.

Today such places have a larger population and more jobs than the cities they surround, and their swath is exemplified by a highway sign Garreau saw in King of Prussia, Pennsylvania: MALL NEXT FOUR LEFTS.

In a way, these edge cities can trace their origins to earlier generations' civil defense concerns, which gave rise to the sprawl and loop highways like the Capital Beltway that are now being so heavily developed. According to Garreau, in many cases official Washington planned for a violent future by placing important structures away from the center of cities. Having seen the destructiveness of nuclear weapons, postwar city planners, relying on military estimates, built the Capital Beltway at a twelve-mile radius around the city. Their military estimates of the destructiveness of a hundred-megaton bomb hitting the White House would stretch ten miles in every direction, and they wanted the nation's critical infrastructures to continue functioning in case of nuclear war.

"The concept was that every major city had to have not only a route that penetrated the city but routes around the city," explained Francis C. Turner, staff director for the commission that wrote the report that led to legislation creating the Capital Beltway, in a 1986 interview. "So in case a bomb dropped, like in Hiroshima, the military needed a route to go around the city, to bypass it." That is also, Garreau says, why half of the Washington-area offices for the federal government lie outside of the District of Columbia.

While the strategy of sprawl by design may have been effective against total disruption of the government in nuclear attack, this new metropolitan model does not offer the same modicum of reassurance in the face of bioterrorist attack. The splotches of high-rise development do spread the population out more than in the packed-in urban centers, but a well-prepared biological attack from the air could cover a large enough area to encompass cities and their edge cities as well. On page 75 is a chart from the 1993 Office of Technology Assessment report referred to in Chapter 1; it illustrates how an aerial line release of properly prepared anthrax could blanket an area of three hundred square miles.

According to estimates by the World Health Organization in a 1970 report, an anthrax attack releasing fifty kilograms of agent—about one hundred pounds—along a two-kilometer line upwind of a population center of 500,000 could lead to 95,000 deaths and 125,000 people incapacitated. Those chilling figures from a report thirty years old are still cited today, though Dr. Edward M. Eitzen, Jr., of the U.S. Army Medical Research Institute of Infectious Diseases at Fort Detrick, Maryland, says that "these estimates are conservative" when compared to the analysis of other experts.

But attack may not come in the open air, which a number of factors make a risky proposition for any terrorist. Atmospheric conditions are out of the killer's control, and a momentary shift in the direction of the breeze can completely change the course and severity of damage. An overly strong gust could render some agents all but harmless by scattering them, literally, to the winds until they can become buried in soil or killed by sunlight. According to sources cited by Stern, the June 1994 sarin gas attack by Aum Shinrikyo in the Japanese town of Matsumoto—

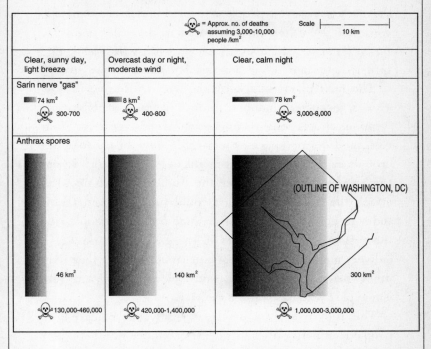

Comparing Lethal Areas of Chemical and Biological Weapons: Delivery by Aircraft as Aerosol Line Source

☠ = Approx. no. of deaths assuming 3,000-10,000 people /km²   Scale ├────┤ 10 km

	Clear, sunny day, light breeze	Overcast day or night, moderate wind	Clear, calm night
Sarin nerve "gas"	74 km² ☠ 300-700	8 km² ☠ 400-800	78 km² ☠ 3,000-8,000
Anthrax spores	46 km² ☠ 130,000-460,000	140 km² ☠ 420,000-1,400,000	(OUTLINE OF WASHINGTON, DC) 300 km² ☠ 1,000,000-3,000,000

Figure shows the lethal areas of single airplane-loads of chemical and biological weapons, assuming a highly efficient line-source delivery of the killing agents. A single airplane delivering chemical or biological weapons can be considerably more lethal than a single missile. For an anthrax attack, the diagram shows how fatalities could vary greatly under three different weather scenarios. In one case, that of an overcast day or night with moderate wind, maximizing the lethal area would require distributing the agent in a 4.5 km by 34 km area, which would not be appropriate for most cities; therefore, the figure assumes a more rectangular distribution, which would still generate a comparable number of casualties.
Source: Office of Technology Assessment, 1993.

which occurred nine months before the Tokyo attack —
was aimed at three judges staying in the area. A sudden
shift in the wind sent the sprayed gas drifting over the surrounding community. The judges fell ill, but they were far
from alone; within the town of 200,000 some one hundred

miles west of Tokyo, scores of people were rendered un-conscious and others were nauseated and feeling ill. "By the next day, seven people had died and more than 200 were ill," according to an account in the 1999 book *The Eleventh Plague*. "Fish in a nearby pond were dead, and birds, dogs, and other animals lay dead in the street."

The next major attack by the cult, as the world now knows, occurred indoors. Indoor release of biological weapons shows some distinct advantages over an outdoor release. Interior spaces themselves now bring together crowds as large as any that might be found on the down-town streets of a city. When the Aum Shinrikyo chose to attack the Tokyo subway, it could have exposed multi-tudes to grave personal harm: some 5 million people ride the system daily. Impurities in the gas and the low-tech delivery method kept the casualties much smaller than they otherwise might have been, said Kyle B. Olson, a terror-ism expert who has studied the cult.

The Tokyo subways are clearly only one place where people routinely gather in such tremendous numbers. Not only do we live in cities, we assemble by the tens of thousands at sporting events, theme parks, tourist attrac-tions, and shopping malls. Many people daily use crowded public transportation to commute to their jobs in gargan-tuan office towers, both in urban centers and in edge cities. These Brobdingnagian buildings are made possible by superefficient, powerful systems that cool or heat the air and completely recirculate it constantly. As we shall see, a terrorist who makes a study of a building's cooling and heating system could find a number of critical spots that would use the building's own high-tech air system to distribute the microbial killers. And such closed spaces—particularly famous shopping malls or institutions like the World Trade Center—could also carry weighty symbolic value.

A major suburban mall in the United States sought me out for advice on how to better prepare against the possibility of bioterrorist attack. Mall officials knew that the facility's sheer size could make it a tempting target, and the symbolic value of its name might even add some attractiveness. I identified the risks I saw inherent in the structure and suggested actions that they might take to better secure and monitor the hundreds of thousands of people who course through in any given week. After the presentation, though, they just shook their heads. "We don't have the money for that," they said. As far as I know, they've never acted on my advice. At least they recognized that they might have a problem; that puts them ahead of other large-scale facilities across the country that still don't have a clue about the risks they face or how to address them.

Asking how terrorists choose their targets is an exercise that, if carried to extremes of analysis, can degenerate into meaninglessness. The terrorism expert is like a meteorologist: able to tell you why it rained in Baltimore and not in Washington after the fact, but unable to predict the precise target with any certainty. No one can say where "Ed" will strike.

And since we're not likely to abandon our cities and edge cities, we have to do all that we can to prepare for the kind of trouble they invite.

4. ORDINARY MADNESS

"Yessssss!"

Wayne couldn't believe it. After years of searching, he had finally found the prize of his collection. He'd put up with this rotten night job at the hospital lab for years now, making it more interesting for himself by taking home things that he liked—things that might turn out to be useful down the road. And now, on the little plate of agar in his hands, was a sample of something he really, really liked: Escherichia coli O157:H7.

He had gotten the sample a week before, when a child had been checked into the hospital with a case of bloody diarrhea.

Wayne never met the kid, of course: the only introduction he ever got to the hospital's patients was the samples of blood, poop, or tissues that made their way down to the lab in the hospital basement. This stool sample had come down and he had done his usual lab workup; he confirmed that the culprit was, as the docs suspected, a strain of E. coli. *"Sounds like somebody shoulda cooked the burgers better," Wayne said to the lab chief when he handed over the results. Wayne didn't know yet whether it was the worst version of* E. coli, *the O157:H7, but he had a good feeling about this one. So he set the plate aside for later. The kid seemed to respond pretty well to treatment and was soon released from the hospital.*

But then the boy had taken a bad turn and had been checked back into the hospital a couple of days ago. He had stopped peeing, and the tests were showing that his kidneys had been seriously damaged by the gravest complication of E. coli *infection, HUS—hemolytic uremic syndrome. They put him on dialysis, but the damage was too severe; Wayne had just received a report stating that after a painful struggle, the child had died.*

This news made him smile. A bug that could cause HUS was a wonderful find, at least in his eyes. This kid was one in a million—or, to be statistically precise, eight in a million. That's how many cases of HUS you might find in the under-fifteen crowd. Wayne had something especially powerful in his hands. Getting a sample of the germ now would be damned hard to do; by the time HUS set in, it took some high-powered lab work to pull out the E. coli. *But Wayne didn't have to, since he still had the original sample right there.*

So he now does what he had done many times before as he built his collection: He twists the screw top off of a "slant," a clear plastic vial about the size of a cigar tube. Then he takes a thin metal scraper and scoops up a bit of one of the spots. At the bottom of the tube sits a yellowish gel, its surface settled at the distinctive slant that gives the tubes their name. Carefully, Wayne spreads some of the goo from the plate on the top of the growth medium and punches some of it under the surface for good measure. Then he quickly closes the top and puts the vial in an incubator to help the culture grow. He waits a day. The following night when his shift is over, he puts the slant in an ice-filled thermos bottle in his lunch box, carries it out past the oblivious hospital security guard, and catches the bus home.

Lots of people joke that their refrigerator looks like a science experiment; Wayne's refrigerator is a science experiment. When he gets home, he retrieves the slant from the ice and dries off the outside of the vial. He pulls out a Sharpie pen and carefully writes "O157:H7" and the date on the tube. Then he opens his freezer and, with pride, places the vial in a frosted test-tube rack that already contains nearly two dozen other vials like it.

Within a few days, Wayne picks up a new part-time day job, one that he doesn't tell his coworkers at the hospital about. It's not that he needs the money. He has plans. The new job is just a few hours a day in the very early morning, prep work at a big centralized kitchen for some of the city's Catholic schools. Thousands of kids get their meals from this place every day. This is the place Wayne needs to be. "Any damned fool can take a gun into a day-care center and shoot up a bunch of kids," he says to himself. "It takes real talent to shoot kids with bugs, though, and not get caught doing it."

Wayne's job is helping the cook whip up the big trays of lasagna and enchiladas and stuffed shells. He spends a little *time, though, ducking around the corner of the big prep kitchen and getting to know Charlie, the baker. Charlie is a talker, and happy to have the company. He sticks to a routine, too; you can count on Charlie to take a bathroom break about thirty minutes after his first cup of coffee in the morning, once he's gotten his big mixers going. So one morning a few months after taking the job, when Wayne sees Charlie head off to the can, Wayne sidles over to one of the mixers; the bowl is blending about seven gallons of buttercream frosting for sheet cakes. Wayne opens the juice bottle he's brought in with him; it contains twelve ounces of E. coli broth that he's prepared in a tabletop incubator at home. With a flick of the wrist, Wayne dumps billions of bacteria into the giant bowl; the blades of the machine quickly blend the straw-colored, sweet-smelling liquid into the swirling goop, leaving no visible trace. He then drops the bottle into the trash chute, washes his hands with the meticulous care of a surgeon, and goes back to his macaroni and cheese.*

A few days later the health aide at Saint Michael's elementary school appears at the director's door. "We have got a big problem, Father," she says. Dozens of children have fallen ill with stomachaches, nausea, and bloody diarrhea. Dozens more haven't come in at all because of illness. "I've called other local schools as well, and they're all seeing the same thing," the nurse says. The director picks up the telephone and calls the local health authorities.

Over the next two weeks, more than two thousand children are sick; the city's hospitals can't handle the sheer number of patients, and some children are sent to nearby towns. There's no sure cure for E. coli infection; antibiotics don't work, and may even make matters worse by causing the bacteria to dump more toxins into the body. All the medical profession can do is provide supportive care and try to ride the illness out, letting the patient's own body be the doctor. Despite the best efforts of doctors and public health officials, four children die of kidney failure and another sixteen are left in intensive care on dialysis machines.

In the meantime, epidemiologists from the state health department have questioned the children, their families, and teachers about the food that they usually eat at school. Because different schools were involved, but all had had food from the same kitchen, and because there were few other possible cases in the community aside from the schoolchildren, the disease detectives are able to identify the school lunch as the most likely cause. Then they go deeper, interviewing hundreds of sick and well children, and determine that the only thing all of them have in common is that every sick child had the frosted cake—and none of the children who skipped the cake got sick. The possibility that that would happen by coincidence is about one in a billion.

The cake itself is an unlikely suspect, since cooking would probably have killed the microbes, so they focus on the frosting, which would be a rich medium for bacterial growth. But because none of the frosting is left, the epidemiologists can't say conclusively that it is the source. The ensuing days are rife with accusations and questions. The local newspaper conducts a high-profile investigation into health code violations at the Catholic schools' central kitchen, finding numerous minor problems. An ambitious young reporter writes a searing reconstruction of events that puts Charlie, as the kitchen employee with direct responsibility for baking the cakes, at the center of the story. The reporter reveals that Charlie has two drunken driving convictions and had once been reprimanded for lateness; much is made of the *fact that he still lives with his mother. The story quotes an epidemiologist saying that most food outbreaks occur because "some people are just too stupid to wash their hands." The school fires Charlie; the local television news reports that the district attorney's office is drawing up criminal negligence charges against him and that the parents have banded together to launch a class-action lawsuit against the diocese.*

Wayne reads the latest newspaper story with a pang of regret. "Sorry, Charlie," he says to himself. "Somebody had to take the fall, and it sure wasn't going to be me." He had quit the job at the school; in the frenzy surrounding the outbreak, no one had even noticed. And since he hadn't had anything to do with bakery work, no one had even questioned him after that first day. The question of his other job at the hospital never came up.

He opens the freezer and looks inside. His eyes fall on another vial marked Shigella. *This was another good sample. A traveler had come back from Asia with a case of the disease, which can cause the same kind of dysentery and kidney problems as* E. coli. *Shigellosis is usually treatable with antibiotics, but Wayne knows that this strain was resistant to just about everything they had in the hospital. He smiles.*

"Yesssss!"

MORE COMMON AND deadly than bacterial meningitis, toxic shock syndrome, and flesh-eating strep combined, foodborne diseases affect tens of millions of people each year in the United States. The Centers for Disease Control and Prevention annually counts some 400 to 500 outbreaks, some involving thousands of individuals, but far more cases go unreported. The most recent estimates of the true number of people sickened each year in the United States is a staggering 76 million. This includes 5,000 deaths and more than 35,000 hospitalizations. We're not even completely sure of all the diseases caused by eating contaminated food. Clearly, lots of illnesses characterized by diarrhea and vomiting have a bad food item as their source. But we continue to find more illnesses like meningitis, blood infections, some forms of arthritis, and even mad cow disease (the human form, in any case, which is called new variant Creutzfeldt-Jakob disease) that can come from eating contaminated food.

Fortunately, none of the deadly, contagious agents like smallpox virus or plague are transmitted via food. It's true that one form of anthrax, gastrointestinal anthrax, can infect someone who eats meat taken from an animal that died of the disease, but it would be very difficult for a terrorist to reproduce the same circumstances. As we just saw with Wayne, however, there's ample room to cause tragedy within the narrow constraints of foodborne pathogens.

Even though the United States is justifiably touted as having one of the safest food supplies in the world, we have witnessed large and often deadly epidemics caused by seemingly "safe" foods. Some of these foods are ice cream, pasteurized milk, apple juice, hot dogs, lettuce, and alfalfa sprouts. Typical germs like *Salmonella*, *Campylobacter*, and *Shigella* bacteria cause most of the epidemics. More recently, an increasing number of the illnesses and epidemics are being caused by more exotic germs like *Escherichia coli* O157:H7, *Listeria monocytogenes*, *Cyclospora cayetanensis*, and caliciviruses. The fact that these latter bugs don't mean much to the average citizen shows how little we know about the causes of foodborne diseases in this country, not only in terms of the germs but even how the contamination of the food happens.

A WEAPONS DELIVERY SYSTEM

Eating, like breathing and drinking, is something we take for granted each day. We would never assume that activities so commonplace, and yet so vital to our survival, would put us in harm's way. But if a potential terrorist wants to introduce a germ into our bodies while bypassing all of our normal immune defenses, then food and water, like inhaled air, are superhighways to our vital organs.

If this sounds unlikely, keep in mind that the only documented examples of bioterrorism in the United States to date involve the intentional contamination of food. The first United States case involving the use of a biologic agent occurred in 1984, when members of the Rajneeshee cult in The Dalles, Oregon, intentionally contaminated salad bars in area restaurants with *Salmonella typhimurium*. More than 750 cases of illness were documented. The motivation of the cult was not entirely clear, but it seems that they were experimenting with this contamination method for later use in that year's November elections. The cult, in an attempt to ensure victory of its own candidate for a local election, wanted to prevent nonmembers from voting. The group did not, however, carry through with its plot—in part because it was apparent their candidate would not win.

A second episode of the use of biologic agents in the United States to intentionally cause disease occurred in a Texas hospital. A disgruntled laboratory employee took cultures of *Shigella dysenteriae* from the hospital laboratory and contaminated muffins and doughnuts placed in the staff break room on a specific day. Twelve of forty-five laboratory staff developed severe, acute diarrheal illness; eight had *S. dysenteriae* isolated from stool and four were hospitalized. Because of the unusual cluster of illnesses among coworkers, an investigation was quickly launched. *S. dysenteriae* was found in uneaten muffins, and genetic fingerprinting analysis showed that it was identical to the *Shigella* bugs found in the workers. The lab employee who committed the act was identified through further investigation; she acknowledged intentionally contaminating the food items.

A skeptic may ask why intentional contamination of food hasn't happened more often if it is so easy. But maybe it has.

One of the most striking facts about the Rajneeshee episode in Oregon is that the involvement of the cult was not established through the epidemiologic investigation of the state and federal public health authorities. As in the fictional case of the cake frosting, public health officials did determine that the salad bar was involved, but no one considered a terrorist as the ultimate source of the outbreak. It was only a year later, during a separate criminal investigation of Rajneeshee members, that their role came out. The notion of an intentional foodborne outbreak does not come naturally to epidemiologists, and if a group like the Rajneeshee doesn't want to announce its attack, authorities might well assume the outbreak is a wholly unintentional result of poor sanitation. Investigators naturally look to people like "Charlie" in the school-kitchen scenario before their suspicions will lead them to "Wayne."

The scenario reflects the way most investigations are conducted: no one determines how the frosting became contaminated, and no one raises the question of intentional contamination. The press and authorities choose a less unsettling explanation—Charlie's incompetence—despite the lack of hard evidence to support it.

Having led one of the elite medical detective groups in the country for twenty-four years, I know that public health officials miss many foodborne outbreaks, even large ones, because reporting and detection are still spotty nationwide. Even the best state and local health departments miss outbreaks. When health departments *can* document an outbreak, they don't always figure out how it happened.

Tracking foodborne disease is growing incredibly complex. Factors that contribute to this new order of food safety include our changing diet, global distributors of food, expansion of commercial food services, and new methods of massive-scale food production. Each of these factors can obscure a possible link to bioterrorism. For example, data compiled by the U.S. Department of Agriculture show that actual dollars spent on food rose from $100 billion annually in 1970 to more than $700 billion annually in 1997. This increase in wealth relative to the cost of food makes the food supply in the United States among the world's cheapest. For many people today, the variety and volume of food have never been greater and food has never been more affordable. In 1950 a typical U.S. grocery store offered approximately 170 food items, most of them canned. Today an average American grocery store has more than 40,000 items on the shelves, with many behemoths exceeding 60,000 items. Imagine trying to confirm the exact source of widely disseminated products that were intentionally contaminated in some distant processing plant or at a foreign farm—a job that could stump the best pathogen detective.

Today almost 45 percent of all food expenditures in the United States are spent on eating food outside of the home. The inclusion of salad bars and nonsmoking areas is currently one of the most important factors by which consumers decide where to eat. Yet as was demonstrated with the Rajneeshee cult, salad bars are extremely vulnerable to intentional contamination by both workers and other patrons. And since salad bar food is not cooked before consumption, there isn't much a customer can do to assure himself of its safety. As the Minnesota Department of Health has demonstrated, such outbreaks are difficult to document, particularly those involving restaurants serving a high number of patrons who blend back into their communities.

Methods of food production and distribution are also contributing to increases in foodborne illness—and possibly to unrecognized biologic terrorism. For example, in September 1994 the Minnesota Department of Health detected an increase in the number of reports of a particular type of *Salmonella* infection. After almost a month of investigation it was determined that Schwan's ice cream, a nationally distributed brand, was responsible for salmonellosis cases throughout the United States. The outbreak was associated with the hauling of unfrozen ice cream mix

in tanker trailers that had been contaminated by carrying raw egg mix on backhaul trips. It was estimated that
225,000 people became ill from eating the ice cream. Despite the fact that this is still the largest documented foodborne outbreak in the nation's history, it took *three months*
before we, this group of elite medical detectives, found it
and were able to take corrective action. It thus becomes
understandable why consumers can't take great comfort
in our nation's current public health disease surveillance
and response capability to detect and respond to a biologic
terror event involving a widely distributed product.

THE ONLY THING ON OUR SIDE

Fortunately, only a very few of the infectious agents that
cause foodborne or waterborne diseases are likely to make
the average person severely ill. While a patient may feel as
if he is going to die as he simultaneously experiences vomiting and diarrhea, death is a relatively rare outcome. Unlike the bacteria and viruses that are typically considered
when one describes bioweapons (such as smallpox and anthrax), those transmitted via food and water are much less
deadly. Taking the standard estimate of 76 million illnesses
and 5,000 deaths due to contaminated food consumption,
one can quickly do the math: only one in approximately
15,000 cases of foodborne disease will end in death. On
the other hand, some foodborne agents are much more
likely to cause severe illness, particularly in young children, immunocompromised persons such as those with
AIDS or cancer, or the elderly. *E. coli* O157:H7 in young
children and the elderly is a good example. Up to 3 percent

of young children who get *E. coli* O157:H7 will go on to develop HUS, like the child in the scenario.

Unintentional outbreaks already cause fear and discomfort; the knowledge that someone has intentionally poisoned people will create panic in the community. Public health officials must always be willing to consider that a foodborne disease outbreak could be the result of a terrorist act. Imagine government officials trying to convince the population that the food supply is safe when in fact the next attack is being delivered to grocery stores or restaurants as the pronouncement is being made. The high level of consumer confidence that the American food supply is the safest in the world—confidence created and reinforced through billions of dollars spent on food safety, inspection, and regulation—would be shattered.

5. TOOLS OF THE TRADE

Yuri shambles through the mall, seven stories of awesome luxury. There are huge gold ribbons, brass ornaments, and twinkling lights everywhere, green and red wreaths with huge pinecones. Bright instrumental Christmas tunes fill the air, along with the smells of caramel corn and cinnamon buns and Californian gourmet pizza. With more than 125 stores, from the grand F.A.O. Schwarz toy emporium to tiny boutiques that sell nothing but fine pens, Chicago's famed Water Tower Place seems to offer the world and all its pleasures.

Yuri hated it. He couldn't get used to the American way of extravagance; the Christmas spirit to him was just a gaudy orgy of consumerism.

His disgust had come as a surprise. Back in the Soviet Union—back when there was a Soviet Union—he would have imagined loving such a scene. It would have come to him only in his most generous fantasies, the daydreams he would slip into to avoid the grimness of his days. But then, his life had been so full of disappointments that if the United States lived up to his expectations, the surprise of it would probably have been enough to kill him.

It had been that way ever since his childhood, his brilliance unrecognized from the start. His aunt, an angry alcoholic who raised him after his parents died, let him know how much she resented the obligation he embodied—that is, when she spoke to him at all. Most often as he was growing up, she reminded him with the back of her hand. His fascination with science couldn't get him far in his remote, rural town; his bookishness simply got him teased and occasionally beaten by the coarse bumpkins around him. Even when he qualified for university training in biological sciences, he was shuttled into the scientific bureaucracy of the state. He could culture bacteria and viruses that others had tried and failed to grow innumerable times, but he had no skill at career advancement. He had waited for them to recognize his gift—but he had waited a long, long time, and found that credit for whatever he did well was taken by others.

Yuri had been one of thousands of workers at Vector, Biopreparat's grimy, hazardous death complex near Koltsovo, Novosibirsk. Vector, one of forty Biopreparat sites devoted to developing biological weapons, became the official home of Russia's stockpile of the smallpox virus after the 1991 fall of the

Communists. When Yuri thought of the place, even now, he would curse softly under his breath. Vector workers grimly joked that the Siberian winds that swept across the open, barren fields surrounding the facility were lazy: they didn't take the trouble to go around you, they just cut right through. Vector's thirty buildings were ringed with electrified fencing, and Yuri often wondered whether the idea was to keep the curious out or to keep the horrors in.

Yuri was miserable at Vector, too, of course, and his anger grew to a pit of bitterness he seemed to carry with him in his stomach. Why couldn't they see that he was entitled to more, capable of more? Behind his back, the bosses thought of Yuri as a neudachnik, a loser. He seemed to them an educated fool, an agonizingly slow worker whose silence bespoke an insufferable arrogance. Although he was clumsy and awkward by nature, he became graceful and sure behind a lab bench; he had what scientists call good hands. Yet this gift—an ability to perform microbiological and mechanical miracles with his ten fingers—brought him no glory. There was no use for his delicacy, since techniques at the plant were rough, and production volumes were unimaginably large. They were in the business of brewing up death-dealing bugs in massive batches: payloads for bombs and missiles. Some of Yuri's coworkers prayed that the devices they made would never need to be used; others walled off the horrors they manufactured from their conscious lives. It was just a job, they would tell themselves, and the weapons— like the nukes—were probably never going to be launched. They saw themselves as being in the deterrence business; Yuri saw them as hypocrites.

Yuri was not one of the worriers, nor was he one of the hypocrites. The idea that these live killing machines might someday *be used to wipe out people by the thousands was a secret source of pleasure to him. Humanity, he felt, was overrated; every day of his life had proved it. If they couldn't recognize his gifts, the bastards, the* sukin syn, *then they could all go to hell. And he knew the means to send them there—had immersed himself in the technologies of death.*

So he had built the thing he thought of as his glorious device on his own. It was, he believed, the perfect tool for a targeted attack. He had worked out the details of the tiny delivery system in his head, never committing anything to paper. Then he had built it with the same painstaking obsessiveness that earned him low evaluations in his work. It took the better part of two years to craft it, spending his nights hunched over tiny valves he had machined with hand tools and nozzles of brightly polished brass. He smashed it and started again three times before he felt that he had made it perfectly reliable. When it was done, he had tried to show it to his immediate superior, Manarov. The imperious scientific bureaucrat had brushed him aside without even taking a look. Yuri had neither the training nor the rank to be able to credibly say that he had solved the daunting problems of creating such a small biowar device on his own.

Manarov's death later that year from an anthrax infection was quietly covered up by the authorities; he was said to have eaten tainted meat. His colleagues wondered how the usually punctilious Manarov could possibly have allowed himself to become infected. Why wasn't he vaccinated? Yuri didn't participate in these discussions, though his mood and the quality of his work improved for a bit.

Then Gorbachev fell, and Yeltsin was forced to tell the world about the country's biological warfare apparatus. He started shutting down the plants. He said he was working toward world peace, of course, but everyone at Biopreparat knew that the nation was out of cash. Yuri had been paid only sporadically as the old Soviet Union had crumpled around him; he applied for a visa to enter the United States, the land of opportunity, and was pleasantly surprised to hear that he had been approved.

When his aunt heard he was leaving, she sent him the first letter he had received since leaving home for the university. He let it sit for a few days before opening it, but finally was too curious and unsealed the envelope. It was a paranoid screed that accused him of ruining her life; she closed with a demand that he send money from America to pay her back for the trouble he caused her. He sent her a curt letter with a liter bottle of good vodka, and smiled soon after when he called his hometown and confirmed that she had died suddenly.

Just before leaving Russia, he packed away his lovely gadget inside the hollowed-out works of a portable radio, and for good measure smuggled out a real treasure: a bit of smallpox. Although the free world thought that the smallpox virus was under tight security, he found it rather easy to get during those days of "slash-and-burn" chaos at Vector. He was certain that his little sample could be worth more than all the gold of the czars to the right buyer.

"Liberating" it had been remarkably easy. Infection wasn't a fear, since his vaccinations were current, but solving the problem of possible detection kept him awake at night. He machined three tiny metal vials that looked like tenpenny nails without the heads; he had to use tweezers to screw the tops on. They were small enough to fit inside the workings of a fountain pen, leaving enough room for a specially made cartridge with just enough ink for him to be able to write with it if challenged. But

no one notices a pen; he sped through customs. Nothing to declare, thank you.

Yuri's immigration forms listed him as a scientist with long experience in molecular biology and biotechnology, all true. He just wasn't 100 percent honest about the things he'd done with that specialized knowledge back home. Cast just right, his deadly expertise would be portable enough to secure him a pleasant job in academe, or even in a biotech start-up that could make him a multimillionaire. He had heard that other Biopreparat scientists had done the same.

Things didn't work out that way. Yuri's command of English turned out to be much less impressive to prospective American employers than he had hoped, and he still couldn't seem to get people to take him seriously. Americans, it turned out, were just as stupid as Russians. He found a lowly bench job in a Chicago pharmaceutical company, his duties limited to checking samples from each batch of medications to ensure that the company's wares were meeting quality standards. It was just as tedious as his old job in Russia, without even the thrill of danger to keep him awake. The job was considered so unimportant that his lab was tucked away next to the storage rooms for old equipment. That was fine with Yuri, though: he didn't want a lot of visitors. And he was surprised to find that some of the equipment that the company had discarded as obsolete was perfectly suitable for his purposes—was, in fact, more modern than the junk he'd worked with in Russia to grow and harvest his smallpox. What the storage rooms didn't have, equipment supply catalogs did. The land of plenty! Ever the tinkerer, he improved upon his device by the week.

He tried watching television to improve his English but found it vulgar and depressing—until one night, after a few hungry months, he saw a movie that gave him an idea. The show raised the prospect that foreign agents might sell their expertise in nuclear technology and biological warfare to the highest bidder. Yuri decided it was time to make capitalism work for him, and vowed to sell his knowledge, his device, and his deadly "Russian gold" together.

Yuri had no objection to such a liaison—a job's a job, after all—but had no idea about how to contact international terrorists. Where did one find these high bidders? A life spent in the Soviet system had left him no better prepared for this facet of capitalism than it had any other. The Japanese cult Aum Shinrikyo didn't exactly advertise its phone number. He had tried to make contact with an Arab organization in Chicago, but they didn't actually seem interested in terrorism. He could only get as far as a functionary whose English wasn't much better than his own and who seemed too stupid to appreciate the beauty of his device. Yuri was afraid to draw attention to himself by doing more.

Then one day as he sat eating beans from a can, slouched in the vinyl recliner he'd found out on the street, Yuri heard a soft rustle by the door. At first he thought nothing of it; by the time he walked over to check, and discovered that somebody had slipped an envelope under the door, the visitor who had done so was long gone.

The letter was brief.

"We understand that you have expertise and materials that interest us greatly," it began. "If you can prove your abilities to us, you could become very rich."

Whoever wrote the letter knew something about him, but did they truly know the power at his command? Could they really appreciate his expertise? "Details will follow," it concluded.

Had his dreams been answered? Yuri broke into a cold sweat. He had no idea who had dropped off the note. It could have been a crazy person. No, that was impossible, Yuri realized; the stranger knew too much about him and appeared to recognize his potential. On the other hand, this could be a trap set by the American FBI, trying to draw him into the light. He stood there by the door, paralyzed with indecision and fear for uncounted minutes. When he finally did move, he went back to the recliner, throwing the beans into the trash on the way.

"To hell with this," he said aloud. If it was a trap, so be it. Prison could hardly be worse than this pathetic excuse for a life. And if his new, er, employers were genuine, then he would soon be able to afford far more than beans. More important, he would be able to use his ever-improving device. He would be able to kill these sukin syn *around him—by the thousands. Perhaps, even, by the millions.*

He arrives at the mall just as the stores are opening, though shoppers are already streaming into the place. He makes his way through the bedlam of the mall. His pocket holds the device, disguised as a thermostat box. Inside are two vials with small, strawlike hoses emerging from the screw tops. One contains two tablespoons of clear fluid, the propellant. The other contains three tablespoons of cloudy fluid. A thin hose connects them to a micro-aerosolizer no bigger than a pack of gum. It is better than anything he could have built himself; he couldn't believe his eyes when he saw it in the catalog at work. He was especially proud of that part of the system: it could transform the small amount of fluid into a gaseous material that would permeate all seven floors of the mall with millions of "infectious hits" in a matter of hours. The motor had taken months of experimenting before he got it right, but the power supply was simple: a small store-bought camcorder battery.

Many things had happened since the dim afternoon when Yuri got his first note. After two days, he received another note, this time demanding a meeting. The instructions were concise but clear.

Yuri had met his contact at the appointed time in the Field Museum of Natural History, by Bushman, a stuffed lowland gorilla that had once lived at Lincoln Park Zoo. The man knew Yuri by sight, of course, and simply nodded to him. Yuri was so frightened he could barely nod in return. As the two men walked through exhibits of dinosaur bones and gemstones, the stranger asked increasingly detailed questions of Yuri about what he had made and what he could make. Yuri, in return, asked nothing. He could not tell the man's nationality; he spoke in what seemed to be completely unaccented English.

After an hour's aimless walking and incessant grilling, the man seemed satisfied that Yuri could deliver. Yuri didn't understand yet what the man wanted him to do, but it scarcely mattered.

"Be ready," he ordered. "You will not see me again. Your next message will tell you where to deliver your work; you will have seventy-two hours' notice, no more. Until then, make certain you have grown enough of your product to perform its appointed task. Have it ready to go within a week. In your old job, you received the vaccinations, yes?"

"Of course," Yuri responded. "Otherwise I could never have survived."

"Good," the man replied. "Should you be successful at this little errand, you will receive $50,000. Consider it the first payment in a long and lucrative relationship." He shook Yuri's hand and then slipped away into a crowd of tourists.

Yuri received the next message on the Tuesday before Thanksgiving. The holiday meant nothing to him, except that he didn't have to go to work. In the weeks before the note arrived, his chores had kept him busy: growing the virus within the confines of his apartment was exacting work. He had to take many steps to be certain no virus could escape his homemade laboratory. Even one case of smallpox in a neighbor might tip off officials to the location of his simply elegant bioweapon plant. The cast-off equipment from work was put to good use, and it proved easy enough to incubate his smallpox culture in eggs from a local hatchery. The dried smallpox virus he had carried into America was growing back as potent as ever.

When he learned of the site he was to hit, a rare smile crept across his lips. It was so beautiful, this plan: a perfect way to maximize the strike. A shopping mall during the beginning of the Christmas shopping rush.

And now it was finally happening. As he makes his way up an escalator, Yuri fingers the one external control on the sleek box: the timer. As he does, however, his finger accidentally flicks the timer switch. "Idiot!" he thinks. By the time he gets off the escalator, he is drenched in sweat. The timer was set to go off in half an hour, and he hadn't designed a cutoff switch. He has to act quickly.

He plunges into a crowd of people heading to the left. Along a narrow corridor, past a row of shops and just before the second-floor elevators, is a return vent responsible for circulating hot and cool air throughout the mall. Although the mall's heating system will destroy some of the virus, more than enough will survive to spread quickly throughout the seven-story structure.

Seeing no security guards, he attaches the device to the wall with rapid-fasten Velcro strips in one practiced motion. In twenty-five minutes a thin mist, invisible and odorless, will spill out of the aerosolizer, far too small for anyone to detect. The particles will be slightly bigger than one one-millionth of a meter— the size of the perfume particles hanging in the air at the Lord & Taylor cosmetics counter. Thousands of people in the mall will inhale the virus into their lungs without even knowing it. The virus, he thinks with admiration, is a truly beautiful weapon, more elegantly designed than even he could come up with.

Right on time, the nearly silent aerosolizer does its job. Armando Gutierrez, a Chicago schoolteacher looking for a Christmas present for his wife, notices nothing as the particles make their way into his lungs.

Neither does Kristin Schafer, who drove in from Milwaukee for the day just to hit the luxurious mall. She's brought her daughters with her, a family tradition: shopping, and then a performance of the Nutcracker before heading home. Nothing ever happens in Milwaukee. "Come on, slowpokes!" she calls to them as they dawdle at a distracting storefront.

It's virus roulette out there. Bill Reynolds, a shoe salesman at one of the mall stores, doesn't breathe in any viral particles. The only thing on his mind is the misery of working his way through the Christmas shopping throng. "Christmas is going to kill me," he mutters.

THE FINAL SCENARIO that takes up the remaining chapters of this book describes a smallpox attack against a major city. In addition to showing how the horror spreads from customers and employees of a shopping mall to become a major international catastrophe, our scenario illustrates some of the most urgent issues within the national debate over biologic terrorism. We begin by focusing on the tools of the bioterrorist trade, and ask just how easy it would be for a lone terrorist, a terrorist organization, or a state-sponsored group to acquire and use the tools necessary to mount an attack.

A bioterrorist needs two things: the germs and the means of distributing them. Each of the two elements of a biological weapon presents challenges to would-be terrorists and to the law enforcement officers who would try to detect or track them.

The difficulties inherent in creating these weapons have led some observers into overconfidence and even a touch of cynicism. The doubters generally argue that making biological weapons is too difficult to constitute a credible threat. The failed attempts by Aum Shinrikyo to deploy its anthrax weapons, they say, shows that biological terrorism is out of reach for all but the most advanced, best-funded groups. Columnist Daniel S. Greenberg even wrote in a 1999 *Washington Post* article that "there's a whiff of hysteria-fanning and budget opportunism in the scary scenarios of the saviors."

Some of the foremost experts in biological weaponry have said things that seem to support that point of view. "I know how to make a weapon," says William Patrick, who developed biological weapons for the United States during the 1960s. "Ken Alibek knows how to make a weapon. I don't think our domestic terrorists have the capability to make a weapon yet . . . it takes some doing, it really does."

Alibek agrees—up to a point. In *Biohazard* he writes

that making the kinds of biological weaponry that he and his staff developed is not a trivial matter, even if you can cultivate a dangerous pathogen. "The most virulent culture in a test tube is useless as an offensive weapon until it has been put through a process that gives it stability and predictability," he writes. "The manufacturing technique is, in a sense, the real weapon, and it is harder to develop than individual agents."

But Alibek and Patrick represent the biological weaponry of the Cold War. Their weapons were battlefield weapons and required the kind of investment and expertise that went into full-scale warfare. The world has changed, and so have its dangers. The high-tech missiles and bombs that Patrick and Alibek helped develop in the heat of the Cold War represent only one kind of attack mechanism, one that rogue nations and terrorists are unlikely to use even if they could. Such missiles could be traced back to their launch site; the brutal retaliation that would almost certainly follow is something terrorists can and would avoid. Terrorists are far more likely to pursue less expensive technology that would still exploit recent advances in benign uses of biomedical and aerosol technologies and take advantage of the incredible destructive power of the biologic agents without drawing attention to their source.

That's why Alibek told members of Congress in October 1999 that the threat of bioterrorism weapons is very real: "Although the most sophisticated and effective versions require considerable equipment and scientific expertise, primitive versions can be produced in a small area with minimal equipment by someone with limited training." They would be, he said, "relatively inexpensive and easy to produce."

The distinction is vitally important. The missile assault that the United States was preparing in the 1960s, and which the Soviets labored on until 1992, is indeed outside the range of even the best-funded terrorist group. The technological wizardry required for turning germs into weapons and then creating delivery systems that can get them to their targets in a still-deadly state takes the kind of all-out development effort that only nationally funded programs can mount.

By contrast, a low-tech outdoor release of pathogens by plane or spray truck—the kind of truck commonly seen on the streets of New York City spraying the pesticide malathion during the summer 1999 West Nile fever outbreak—is far easier to accomplish and could infect multitudes. No more difficult would be an indoor release of aerosol particles within an airplane, airport, or shopping mall, all to devastating effect, especially if a highly contagious agent such as smallpox is involved. Today's megamalls host enough people to fill a small city each day. Even from a starting point as small as the few hundred passengers on a commercial airliner, the ripple effect of infection would be more deadly than an urban bomb blast, for as the victims dispersed, each would be the potential carrier of a new epidemic.

What the skeptics fail to acknowledge is that the tools of microbiology and medicine have made obtaining and cultivating pathogens *and* building basic weapons that use them simpler, cheaper, and more widespread than ever before. "Even groups with modest finances and basic training in biology and engineering could develop, should they wish, an effective weapon at little cost," writes civilian biodefense expert D. A. Henderson.

We have already seen the proliferation of biomedical knowledge that would make it possible for a terrorist to learn how to cultivate lethal biological agents. Equally troubling is the relative ease with which such activities can proceed undetected: anyone hoping to track the proliferation of technologies for growing the worst bugs must deal with the fact that they are precisely the same technologies that are used every day for beneficial purposes in labs around the world. Many technologies have such "dual uses," but the problem is especially vexing in the burgeoning field of biomedical research, where scientists often need to study the very same pathogens that can be used for terrorism.

Pulitzer Prize–winning author Laurie Garrett explored the especially difficult dual-use issues in biomedicine in a bracing 1998 investigation for *Newsday*. No legitimate scientist, for example, would ever need to enrich a large enough quantity of uranium to produce weapons-grade plutonium. But bioweapons production, she points out, "looks just like medical research or vaccine development — indeed, to make a vaccine against such things as anthrax, or salmonella, one must have supplies of the organisms on hand. Minute amounts of bioagents can nestle innocently in a lab freezer alongside legitimate biological substances. And yet, that tiny tube may contain enough biological material to kill massive numbers of people."

Garrett spoke with Dr. Raymond Cypress, director of the American Type Culture Collection (ATCC), a Virginia-based repository of thousands of tissues, bacteria, and viruses that can be ordered by mail. The ATCC is a critical tool for the biological sciences, but gained notoriety when Larry Wayne Harris used the repository to order samples of *Yersinia pestis* in 1995, and was found to have shipped samples of anthrax to a lab in Iraq. ATCC, while huge, is not alone, Garrett writes: "Though ATCC is the largest microbe supplier in the United States—one whose Web site alone receives more than 100,000 shipment requests each month—there are countless other ways scientists, or terrorists, can obtain pathogens.

"For example, 27 research laboratories in America recently have published work on *Yersinia pestis*, which causes plague, 'but only four got cultures from us. So where did the rest come from?' Cypress asked."

The answer can be found in medical libraries and labs around the globe. The World Directory of Collections of Cultures and Microorganisms lists some 450 such repositories worldwide. None, of course, sell smallpox. But at the time Garrett wrote her article, more than 50 of the repositories sold anthrax, 64 would sell the organism that causes typhoid fever, 34 offered the bacteria that produce botulinum toxin. And 18 repositories, located in fifteen countries, traded in plague bacteria. These businesses— many of which take orders over the Internet—can be found in more than sixty nations, including China, Bulgaria, Turkey, Argentina, and Iran.

Trying to assess which researchers are ordering virulent pathogens for legitimate purposes and which are not is one of the most daunting challenges for law enforcement. Overbroad restrictions on shipments could restrict academic freedom and impede scientific and medical progress. And how does a professor determine whether a possible graduate student or laboratory assistant is a potential terrorist risk?

Some agents are harder to find and produce than others. Smallpox, as Dr. Peters of the CDC notes, is relatively easy to grow once you get it, but acquiring smallpox would be difficult. Yet tragically, the reemergence of smallpox—the worst of all scenarios—would not be impossible. Although the official stockpiles of smallpox are limited to the Centers for Disease Control and Prevention in Atlanta and the Russian State Research Center of Virology and Biotechnology (known as Vector), the scientific and intelligence communities now realize that the "two-lab repository" story is a myth, and a dangerous one. Alibek, the former high-ranking official of the Soviet biological weapons program, said that smallpox didn't stay in the freezer—and that it is just one of the legacies of the Soviet biological weapons program.

And remember, Russia is not the only nation that has secured and produced biological weapons, including smallpox. Alibek told me point-blank: "We make a very serious error if we assume that only the U.S. and Soviet Union have smallpox. . . . When we read WHO saying this virus is eradicated, that may be true from their point of view, but not from the biodefense point of view. At least ten to twelve countries have this agent. I have one hundred percent confidence that North Korea has it."

"We still don't know which countries are doing what with what," D. A. Henderson says, adding that Saddam

Hussein's son-in-law's exposure of the Iraqi biologic warfare apparatus "only illustrated how difficult it is, and would be, to identify what's going on."

For Henderson, the man WHO put in charge of eradicating smallpox, the revelations that the Soviets had been producing smallpox and other lethal agents was agonizing. "I feel we've been betrayed totally by the Soviets in a way that is unconscionable," Henderson says today. He has recently seen documents purporting to show that the labs have had smallpox for only a relatively short period, since 1994. "It was absolute bullshit," he says. "You know this is not true." Alibek, he notes, says the samples of smallpox were obtained in 1980.

Henderson's anger is understandable. Wiping out smallpox, a project that involved the labor of 150,000 people over an eleven-year period, stands as one of the greatest achievements in human history. No single act has prevented more death and suffering. When the program began in 1967, there were still 10 million cases worldwide and 2 million deaths a year; hundreds of millions had died in this century alone. I was saddened to see that Henderson's feat was not recognized when the journalists and pundits tallied up the great achievements of the century in December 1999.

Henderson, a large man, powerfully built at sixty-two, and with a full head of gray hair, was a reluctant crusader for bioterrorism preparedness. I pleaded with him to go public with his concerns about the reemergence of smallpox. After two years, I finally got him to agree to address the Infectious Diseases Society of America in December 1998. He opened his presentation with a qualification; the last one I've ever heard him give. "Until recently, I had felt it unwise to publicize the subject because of concern that it might entice someone to undertake dangerous, perhaps catastrophic experiments," he said. "However, events of the past twelve to eighteen months have made it clear that likely perpetrators already envisage every agenda one could possibly imagine." People in the audience were spellbound by the talk, which many of us thought would spur policy makers to push for rapid and effective action. I've seen several similar opportunities wasted since then, though, and now see that I was naïve to think that one speech might change the world. It did, however, give encouragement to those who do recognize the risk. More important, it helped bring this giant of public health—the ultimate authority on this global scourge—back into the fight.

Henderson and others are now speaking out in part because we've heard from people who have visited the Vector plant how easily some of the Soviet smallpox stockpile could well have walked out of the facilities that once made it. After one 1998 visit to the Vector plant, Dr. Peter Jahrling, an internationally known senior scientist at the army's Medical Research Institute of Infectious Diseases and a hero of *The Hot Zone*, said, "There's no doubt in my mind that the smallpox sample is not secure. I saw the site. The only apparent security was one pimply-faced kid who looked about fourteen with a Kalashnikov rifle."

The virus, the knowledge of how to cultivate and use it, and the people who developed the technologies could all be up for grabs. Tens of thousands of scientists working in the Biopreparat system are now cut off from the profession they trained for, and many are victims of the rollercoaster vicissitudes of the Russian economy.

Like many experts, Henderson believes that the sorry state of the Russian economy means that many of the scientists who had worked on biological warfare have put their expertise on the block. "You've got a group of scientists who are in deep financial straits." Is such a scientist struggling to support his family "apt to do something extreme?" "Very possibly," he says.

The New York Times's Judith Miller reported in December 1998 that the Iranian government had been working to recruit bioweaponeers. "One Russian germ scientist recalls meeting an Iranian recruiter who 'seemed to have visited most of the labs and institutes in the area.'" Another Russian bioscientist recalls visiting Tehran in 1997 and bumping into three former colleagues. The *Times* reported that a recruiter had been sent directly to Moscow by the Iranian government.

Just one of the scientists, Alibek suggests, would be "a bargain at any price," since "the information he could provide would save months, perhaps years, of costly scientific research for any nation interested in developing, or improving, a biological warfare program." He admits that it is impossible to know how many—if any—have answered the call, but says he heard from "a former colleague, now the director of a Biopreparat institute," that five of the institute's scientists are in Iran.

"I know of about twenty scientists who formerly worked for the Soviet biological weapons program and who now live in the U.S. This indicates to me that it has been relatively easy for these experts to leave Russia, and if twenty of them are in the U.S., undoubtedly a number of them are in other countries as well," Alibek testified before Congress in 1998.

Alibek showed lawmakers a "flier advertising the wares of a company called 'BIOEFFECT Ltd.,' with offices in Moscow and Vienna. . . . The flier offers recombinant *Francisella tularensis* bacteria with altered virulence genes. Ostensibly, these organisms are being offered for vaccine production; the flier also notes that they can be used as genetic recipients and to create recombinant microorganisms of biologically active agents. The authors of the flier also express willingness to form cooperative ventures to which they will contribute their genetic engineering knowledge. It is clear from this flier that the scientists of 'BIOEFFECT Ltd.' are willing to sell their genetic engineering knowledge to anyone."

Knowledge is every bit as dangerous as the pathogens. Lawyers say, "You can't unring a bell," which means that once something is said it can't be unsaid. Alibek says that microbes, the instructions for cultivating them and for making them into weapons, could be smuggled out easily—and were—because they take up so little space. Alibek notes that the extensive and exquisitely detailed instructions for the cultivation of killer bugs and the weapons to deliver them—twelve volumes for smallpox alone—were placed on microfilm, making them highly portable.

One U.N. official told me with frustration that smuggling biological weapons or the information for their use would be trivially easy compared to other goods that have been successfully moved out of Russia. "In 1995 Russian missile guidance systems for submarines found their way quite happily to Iraq," he explained; how much easier would it be to "smuggle out virus cultures which could even be hidden in fountain pens"?

William Patrick has done it plenty of times—to see if it could be done, and to prove it. For some time now, he has carried with him a vial containing 7.5 grams of a "simulant" for anthrax: a bacterial culture that looks exactly like the killer bug under a microscope but that is harmless. He made headlines in March 1999 when he pulled out the vial at a hearing of the House Permanent Select Committee on Intelligence. "I've been through all the major airports, and the security systems of the State Department, the Pentagon, even the CIA, and nobody has stopped me . . . seven and a half grams would take care of the Rayburn Building and all the people in it," he said.

A delivery system for biological agents can be as simple as the juice bottle envisioned in Chapter 4, or a vial of dried anthrax spores thrown against a wall to break and release the deadly agent. But it can be much more. Just as the broad expansion of biomedical training has created a vast population of people capable of cultivating dangerous microorganisms, advances in other fields have created rich possibilities for dual-use technologies that could be turned to the creation of inexpensive biological terror weapons. And you don't have to go to Russia to figure out how to make them.

Let's take as an example one of the aspects of "weaponizing" biological agents that is often held out by the bioterrorism skeptics as one of the most daunting: producing the kind of wet or dry aerosol that will effectively distribute a bug. Again, there are voices that downplay the possibility of this technology falling into the wrong hands. "The capability to disperse microbes and toxins over a wide area as an inhalable aerosol—the form best suited for inflicting mass casualties—requires a delivery system whose development would outstrip the technical capabilities of all but the most sophisticated terrorists," write authors Jonathan B. Tucker and Amy Sands of the CBW Nonproliferation Project at the Center for Nonproliferation Studies of the Monterey Institute of International Studies in Monterey, California. They cite the need to produce the right kind of anthrax cloud, with particles of the one-to-five-micron size that can be absorbed within the lungs and will hang in the air, as a key stumbling block.

But experts in the burgeoning field of aerosol technology disagree, saying that the achievements of their discipline would make aerosol distribution simple—and that those technologies have been well described in the scientific literature.

To explore the availability of equipment to disperse wet and dry aerosols, I went to the biggest dual-use technology of them all: the Internet. The global communications network has become a forum for scientific exchange, a marketplace for all manner of equipment, and a communications medium for the best and the worst of humanity. Dozens of Websites offer information on new and used crop-dusting planes and equipment that can be fitted to almost any plane or even trucks. Most of the equipment that can be found on those sites produces a highly controlled wet spray, with nozzles that can set the droplet size precisely. Farmers (and environmental regulators) prefer wet aerosols to equipment that disperses dry powders, because

the wet sprayers offer the user greater control over placement of the spray; powders tend to drift, spreading a farm-
er's pesticide far afield and raising environmental risk. Such
systems would not be the most effective way to distribute a
dry agent like anthrax, but pathogens could be sprayed in
small droplet form to potentially devastating effect. A quick
call to the toll-free number for a state university's agricul-
tural extension service (listed, naturally, on its Website) re-
vealed that powder dispersal systems, while less popular
than wet systems, are still available. One Website even provides a handy guide to the area one could expect to cover
using various particle sizes, wet and dry—from thousand-
micron particles that travel only four feet to half-micron
particles capable of drifting almost four hundred miles.

Not only is information about aerosol technology widely
disseminated on-line, the products themselves can be purchased via the Internet, either new from on-line stores or
secondhand from on-line swap and auction sites.

Specialists in the aerosol field say their techniques have
advanced greatly and costs have come down thanks to
broad use in agriculture, industry, and research. "What
we can do now with aerosols compared to a can of Right
Guard is like the old computers of twenty years ago compared to the Pentium II laptops of today," says one leading
expert in aerosol particle technology who has served as
my mentor in understanding it. Industries ranging from
health care to photocopiers have mastered the technologies of spraying to produce machinery that delivers particles of precise size and quality. And, as with the biological
sciences, the equipment itself is easy to find and portable.
Aerosols can be blown with devices the size of a pack of

gum. Three tablespoons of liquid containing more than 300 million smallpox viral particles and propellant will fit into a device just as described in the scenario at the beginning of this chapter. A total delivery system could weigh as little as five pounds.

The technologies for some weapons are as close as the corner pharmacy, says Dr. David Pui, director of the particle technology laboratory and a professor at the mechanical engineering department of the University of Minnesota. "Actually, there are some types of medical nebulizer that people can buy in a drugstore . . . for just a few dollars, you can put this biological material in a suspension form and spray it—it's really quite effective." Pui has done extensive work in distributing bio-aerosols in the laboratory in just that way for the mundane purpose of testing the filters in air conditioners. "We use harmless microorganisms," Pui says, but "in bioterrorism, that may not be the case." The drugstore variety can be the size of a flower vase, but some can be had that fit in the palm of the hand.

Other handheld dry-powder devices, such as inhalers, could be used to disperse dry agents such as anthrax as well. "The powders used in medications," he notes, are "basically the same size" as the bacteria that would be used in bioterrorism. Converting an existing device to terrorist use by loading a cartridge with a biological agent "really is very, very simple," he insists. "A person can buy an inhaler from a drugstore and learn how to do it . . . we have been doing this type of thing all the time" in the lab, he says.

"Any of my graduate students . . . learned how to disperse a particle very, very effectively," Pui says.

Even without the benefit of Pui's training, terrorists could educate themselves on aerosol technology with the resources readily available in technical libraries. One typical academic paper I found with little trouble goes into great detail to show, with diagrams and instructions, the design and properties of the various kinds of nebulizers, spray cans, and other aerosol-generating devices. Once one is armed with a better understanding of aerosol equipment—such as the name for a particular type of nebulizer—navigating the World Wide Web to find supply houses from which each piece of equipment can be purchased, often on-line, becomes even easier.

Privately, some law enforcement officials admit that they have a long way to go before they are competent to recognize the kind of equipment that could be used in a biologic attack—especially the tiny devices that might be disguised, like Yuri's hypothetical thermostat box. "We've got lots of expertise in explosive devices and even methods for mixing chemicals," says a senior FBI regional official. "But I don't know anybody in the FBI who has any expertise in aerosol particle work. They wouldn't really know what to look for when these things occurred."

William Patrick already knows that. His attempts to test the limits of security systems also include the use of equipment that could be used to disperse biological agents. He often carries a rose duster with him—a gardener's tool for dispersing bacterial spores such as the natural pesticide *Bacillus thurigensis*. The device is about twenty-two inches long and made out of metal. "It shows up so beautifully on X ray," he says with a grim laugh. "I cringe when I see it. . . . When you combine the simulants I have with the rose duster, you'd think somebody would say, 'What the hell is this thing you have?'

"Nobody's ever stopped me."

The problem, Patrick says, is that "these people who man the X rays at the security checkpoints, they're doing a very good job of detecting pistols and knives and what appear to be bombs. . . . They don't have a clue as to what to look for in detecting a BW agent. And that's asking a lot of them! A pound of flour could be a pound of flour, or it could be anything else." What will happen if he is ever stopped? "I guess I'll say I'm an expert in roses," Patrick says.

Patrick, who so forcefully makes the point that biological weapons are hard to produce, finds this under-the-table proliferation the key to possible future attacks. "The terrorist group that concerns me is not our homegrown variety—I don't think they're there yet." Instead, he says, he finds the greatest risk in those terrorists who might be backed quietly by a rogue state—one that can marshal the resources to overcome technical hurdles. What's more, he notes, a nation's diplomats "can bring this dry powder in through diplomatic pouch." That is the scenario that worries him, he says; used in a closed environment such as an airplane or building, "they could cause us great grief."

I've known Patrick for years now, and consider him one of the nation's greatest assets against biological terrorism. Unlike a plethora of self-proclaimed experts with no actual expertise, Patrick has been there and built the bombs. He's a lovable, warm guy—the kind of man you'd want as a grandfather to your kids—and as talented, caring, and thoughtful a person as I've met in all my years. And yet, thirty years before Alibek, he created brutal biological weapons of mass destruction for his nation. There was a Cold War on; like Alibek, he believed he was working to match a threat from a resourceful and brilliant enemy. I keep that in mind when people ask me how anyone could do such a terrible thing; how anyone could contemplate creating devices that would agonizingly kill so many. The answer makes me terribly uncomfortable: it could be anyone, even the nicest guy you ever met.

6. LIVING TERROR

Ten days have passed since the silent attack at the Water Tower Place, which distributed millions of doses of smallpox among the more than 100,000 shoppers. The mall draws visitors from all over the region; those exposed return home to dozens of communities in more than ten states.

Thousands of people inhaled the microscopic killers. The brick-shaped virus particles replicate in their hiding place—the lymph cells of the body—for days before symptoms begin. Soon the virus will manufacture tens of thousands of copies of itself, recurring countless times in the secret silence of the body.

The days go by with no sign that anything has occurred. In

the crowds of shoppers at Water Tower Place, no one notices the man who quickly pulls what appears to be a thermostat box off of the wall, leaving only a couple of adhesive-backed Velcro strips behind. That night, a janitor scrapes away the strips and cleans the remaining adhesive off. Any evidence that Yuri was ever there is now gone.

Nearly two weeks after the shopping trip, Armando Gutierrez is feeling terrible. He has a raging headache—unusual in itself, despite his daily dose of noise from his rowdy classes. His lower back aches powerfully; he also has a patchy, rosy rash on his face. Gutierrez, who has not missed a day of school in five years, tells his wife he's going to stay home.

By then, large numbers of patients begin showing up at hospitals with fever, headaches, backaches, and vomiting. Though the patients are extremely ill, emergency room doctors discourage them from checking in. The hospitals are already on "code blue"—an early flu epidemic has every hospital in the area operating at or above full capacity. Patients phone their family doctors' offices to ask for appointments; the nurses try to decide whether to ask them to stay home or make the trip into the already-packed offices. Those who do get seen are given a quick diagnosis of flu and are sent home with instructions to take Motrin, drink plenty of fluids, and rest. Many start arriving at local emergency rooms, where physicians are certain they're seeing the annual flu surge. The worst cases get tested for meningitis and other illnesses, but those results are negative. In the first few days the cases are spread out among so many health care providers that the true size of the outbreak hasn't been grasped.

When Gutierrez's wife gets home from work, Armando is still in bed, moaning. His fever has continued to climb, and he has developed a stomachache as well. She calls the family HMO, but office hours are over. Rather than wait overnight to see someone, Armando asks his wife to get him to the hospital, "Just to check me over," he says.

The ER doc is not impressed—especially since his emergency room is overflowing with patients. "Could be staphylococcal food poisoning," he says, before giving Gutierrez antibiotics and sending him home.

Other patients, including two pregnant women who arrive at the same hospital twelve hours apart, show up with horrible symptoms that stump the doctors entirely. They seem to be developing purplish spots under the skin, a type of hemorrhaging, and then their bodies just go straight to hell. Their respective doctors try to figure out what the disease might be, but none of them have ever seen any form of smallpox, much less the less common hemorrhagic variety from which these women are suffering. The similarities between the two cases would normally be alarming, but because they are in different parts of a large county hospital, no one recognizes a pattern.

The next day, Gutierrez returns to the hospital in a much-weakened state; like many of the most serious patients, he is now delirious and has also been racked with chills and has been vomiting. What had been a pale rash on his face the day before is now turning into small, BB-like blisters. The doctors on call decide among themselves that this must be an adult case of chickenpox—even though that disease usually shows its rash on the torso first, and the schoolteacher's wife insists that her husband had had chickenpox in childhood. Initial tests turn up nothing; the white blood count suggests some kind of viral infection. The doctors put the patient in an isolation room to monitor the course of the "chickenpox" and to

keep him from spreading the disease to other patients (beyond those infected as he sat in the ER waiting room for
three hours before he was seen). In a lucid moment, Gutierrez
turns to his wife and whispers, "It feels like my skin is on
fire."

The next day, Dr. Barbara Bradburn, Cook County's top
infectious disease specialist, examines Gutierrez, noting that
his light rash has gone from blisters to pustules—knobby erup-
tions that are hard to the touch. They cover much of his face in
a horrible display. "Good God!" Bradburn says to herself, recognizing the characteristic smallpox rash. She had just returned
from a meeting of the Infectious Diseases Society of America,
where she had attended a seminar on recognizing the signs of
bioterrorism-related illness; never did she suspect she'd be looking at a living case just days later.

She walks up and down the waiting and exam rooms of the
ER, and then along the packed hallways. The rash is everywhere, on more than a dozen patients, now obvious. The worst
cases have been transformed gruesomely: their blisters are so
plentiful that they have grown together, giving the victims' skin
the eerie look of crepe rubber.

The chief resident keeps asking Bradburn, "What is it?
What's happening?" But she says nothing, turns, and walks
away.

Bradburn steps into her office and closes the door; she doesn't
turn on the light. She crosses to her desk by the glow of the
just-lit streetlights outside her window and sits heavily in her
chair. "The pregnant women . . . it's hemorrhagic smallpox.
They've all got smallpox." Her right hand feels the scar on
her left arm—her own vaccination from childhood, its power to

protect her likely long gone. Fighting the urge to panic, she picks up the phone and calls the city health department.

At first, nothing Bradburn says can convince Dr. Joe Evans, city health director, that she could possibly be right about this. It just couldn't be smallpox, he thinks—she must be mistaken. But as the rapid-fire discussion continues, he finds that he has no scientific or emotional blinders strong enough to convince him that the diagnosis could be anything but smallpox. The nightmare.

After telling Bradburn to keep all patients in the hospital until vaccine can arrive, Evans calls the CDC and reports the crisis. He has just as much trouble convincing them as Bradburn had convincing him. No one wants to believe a caller who suggests that smallpox is back. He puts in an urgent request for vaccine and calls the mayor's office—and is put on hold. He hangs up and calls the police chief to request guards at the hospital.

But before he can get through, he has to stop and ask himself questions he would rather not answer: Who is guarding whom—from whom? Do you not let anyone out of the hospital, or not let anyone in? He realizes that the disease is already all over the city by now and thinks: "Even me."

The ensuing panic unfolds with terrifying speed. After word leaks out in the hospital, efforts to maintain order fall apart as panicked hospital staff and others flee the scene. A distraught emergency room nurse calls a local television station's investigative tip line, and news of the epidemic is broadcast within minutes, sparking an exodus from the city that will spread the disease farther.

By this time, clusters of patients are arriving at the city

hospitals deathly ill, with many dying within forty-eight hours. The city health department calls all of the local hospitals and discovers more than a hundred patients around the city; reports of cases in nearby towns have also come in.

Frightened to go out, afraid to sit at home, residents barrage the hospitals and doctors' offices with telephone calls, demanding vaccinations. Supermarkets are emptied as people rush out, wearing makeshift masks, to lay in provisions for a long siege.

The hospital is overwhelmed—but it's only the first wave. Within a day, patients are flooding all of the local hospitals by the hundreds, their bodies undergoing horrific changes. Each highly contagious patient has already set off a new wave of infection that will exponentially increase the number of victims within another two weeks.

The hospital has no beds, no space, no staff. Even in the best of times, there would not be enough doctors and nurses to help deal with an epidemic of this size. The remaining skeleton crew of doctors and nurses who have already been exposed is left to deal with the growing population of desperately ill people. "Where are the feds?" Bradburn and others ask repeatedly. But despite promises to ride in like the cavalry, no relief is in sight.

The story has taken over the airwaves; the world's attention is fixed on the crisis in Chicago and other cities where cases are beginning to appear. But attention isn't help, which Chicago needs badly. In a conference call with the hospital officials, the governor, and top U.S. public health officials, the Secretary of Health and Human Services offers to erect a tent city on the edge of town to accommodate the overflow of patients—and is taken aback when the health commissioner points out that the hospital's staffing needs already aren't being met. Without anybody with immunity to the disease, vaccine to get people that immunity, or even masks to protect them, no relief program stands a real chance of being implemented. The secretary's comment that the government is "doing everything we can" sends the governor over the edge, and he demands an immediate airlift of vaccine.

"We are doing everything we can," the secretary says again, slowing the words for emphasis. "We're going to need to get approval from the White House to do an airlift like that. We don't know yet how many other cities are experiencing outbreaks, and we're trying to prioritize the uses of the vaccine that we have in-house, getting the diluent and the bifurcated needles. As soon as I hear back from the White House, I'll let you know what we're going to be able to do." She cuts off from the call before the screaming starts.

As the conference call disintegrates, Evans talks with the city's chief of police, who has also been tied into the call.

"The most important thing right now is a quarantine," Evans says. "We don't have a lot of experience with this—you know, New York has still got TB, but this is like the old days of smallpox and polio, and we're going to have to do what they did."

"Well, we'll do what we can, Dr. Evans," the chief says. "Now tell me exactly what we need to do, because it's all new to me, and my officers, too. Don't get me wrong—we've all been part of those boring terrorism planning meetings and training. But none of this was ever covered . . . dammit."

"Okay. When a house is identified with smallpox, you get the patient to treatment and then cordon in the house the family members not yet sick. Placard, yellow tape, whatever. If *we've got the vaccine, we give them the vaccine. If we don't have it yet, they have to stay in, or each person who's picked it up spreads the disease in a new wave. Nobody leaves the home for at least two weeks, until we know if they are infected."*

"Who's going to enforce that? People won't stand for it!"

"You are. By whatever means you've got."

"Dr. Evans, I'm not sure I understand what you're telling me, so I'm going to ask you straight: are you saying that I've got to tell my officers to shoot unarmed civilians because they might be sick? Might be?*"*

"Chief, I don't have any better way to say it. We'll try to explain the importance of this to everyone by television and the newspapers, and hope they understand. But if they don't you've got to stop them—or hundreds of thousands could die."

"My officers won't do it, and I don't have the authority to tell them to. I'm not going to."

"I hope you change your mind—or you're going to kill a lot more people than your officers ever would."

By the end of the first wave, 8,943 people have been diagnosed with the disease in ten states. About 6,000 of them are in Chicago, 500 in Milwaukee, and the rest spread out over the region. A third of them will die; the survivors will bear the disfiguring scars for the rest of their lives. By the time another two weeks pass, the number will leap to 84,000, with 60,000 in Chicago alone. If the disease can't be brought under control, the next wave could infect 300,000, 500,000—maybe even a million.

DR. JOHN BARTLETT walked into the emergency room at Johns Hopkins University hospital just before dawn on

February 13, 1999, to ask the physician in charge about a troubling patient. He described the symptoms to the doctor and asked her what the treatment should be.

The symptoms that Bartlett laid out for the emergency room physician were those of inhalational anthrax. But the doctor—one of only five physicians in the state of Maryland at the time who had taken a special eight-hour training course in responding to bioterrorism—suggested that the most likely diagnosis would be flu, and that the patient would be sent home. Since there was a flu epidemic going on at the time, the symptoms would not raise suspicions, she said.

As you've probably guessed, there was no patient. Bartlett was performing an experiment to test the level of preparedness in our health care system, and found it sadly lacking. He's no crank—he's a professor and chief of the Division of Infectious Diseases at Johns Hopkins University School of Medicine, an internationally recognized infectious diseases researcher, and 1999 president of the Infectious Diseases Society of America. And the early misdiagnosis was only the beginning of a long, depressing day.

From the emergency room, Bartlett went to the hospital's department of radiology and asked the doctor there to examine an X ray he had brought with him, "a classic case of inhalational anthrax," clearly showing the widening of the mediastinum that is rarely found outside of that disease. The radiologist identified the widened mediastinum, but his diagnosis nonetheless "did not include anthrax," Bartlett says.

It got worse. The lead technician in the hospital's laboratory explained that even if the bacterium showed up in a blood sample, it would probably be labeled "*Bacillus* species, a probable contaminant." It would not be subjected to full-scale identification testing until three or more positive samples appeared—and then those tests would take forty-eight hours. Another seventy-two hours would be eaten up by sensitivity testing, the necessary tests to show which antibiotics are effective for treatment or prophylaxis.

Bartlett then did what any medical professional would do in the case of an outbreak as serious as anthrax: he phoned the Maryland Department of Health and Mental Hygiene.

He got a recording. Bartlett says that he left a message stating that "I had a query about bioterrorism, and it was important."

That call was returned three days later. Bartlett explained that the state has set up a response mechanism for bioterrorism events that, in theory, can be activated with a single phone call. "The problem is that I did not know the number. No one else seemed to know the number; it is not on the hospital directory or on 911 listings."

When Bartlett looked into the supplies of drugs necessary to deal with an outbreak, the news was no better. The usual first line of treatment for anthrax victims is ciprofloxacin, a member of the fluoroquinolone family. He found that the city of Baltimore had, at that time, 69,000 capsules of ciprofloxacin and 99,000 capsules of doxycycline, and a number of other fluoroquinolones in area pharmacies as well. That might sound like a lot. But if you consult the medical guidelines for treatment of anthrax that were published in 1999 in the *Journal of the American Medical Association*, you find that the recommended treatment for people sick with anthrax or exposed to it is 2 ciprofloxacin tablets daily for sixty days. That's a total of 120 capsules, which reduces Baltimore's stockpile to treatment for a mere 575 patients—not enough to take care of the sick people in our earlier scenario, much less those exposed who haven't shown symptoms. The second-line treatments don't increase the numbers by much more than that. (The new federal antibiotic stockpile, which has been built up since Bartlett's talk and which I will discuss later, helps address some but not all of the issues.)

Bartlett described his experiment at a conference on bioterrorism; it was later published in the journal *Emerging Infectious Diseases*. As frightening as it was, it received no press coverage that I ever saw.

Bartlett wasn't trying to say that his city was less prepared for disaster than other cities—in fact, his goal was the opposite. He believes that Baltimore has done more than many cities to face up to the prospect of such a disaster; his point was that signals could be missed anywhere, and no city is ready for the onslaught of patients that a large-scale biological terrorism attack would bring on.

You might think that the public health community might have received Bartlett's troubling message as a wake-up call. Not quite: as Bartlett recalls, the general response was that "they were pissed off" at him.

• • •

If smallpox returned as a result of bioterrorist attack, or some other disease was unleashed on an unsuspecting population, how would we figure it out—and how long would it take? We now turn to the true front line of a bioterror attack: doctors, nurses, clinics, and hospitals, where the first signs of trouble would emerge as the outbreak begins to unfold. Diagnosing illness is rarely easy or straightforward. Countless diseases look like the flu at first, and even some very distinctive symptoms can be shared by a number of diseases. When smallpox could still be found naturally in the world, it was often mistaken for other illnesses until a full-blown epidemic of advanced cases confronted physicians.

Even if doctors can come up with a diagnosis, no community is prepared for the onslaught of patients that a successful bioterrorism attack would bring. Despite all of the media attention and government spending on counterterrorism, these are the people who have gotten little help in preparing for the crisis that could someday appear at their doorstep. "Federal, state, and local health agencies play a central role in planning," says Bartlett, "but do not have the facilities or field forces necessary to deal with sick patients and the thousands who need vaccines or antibiotics" for such diseases as anthrax or plague.

Getting more drugs quickly could be especially difficult because of the practices of today's health delivery system and the modern pharmaceutical industry. Like their counterparts in industries ranging from cars to computers, today's hospitals, drugstores, and pharmaceutical companies have slashed the costs of inventory and warehousing by moving to a "just in time" delivery system: little stock is kept on hand until it is actually needed. For example, most hospital pharmacies today receive one or two shipments of pharmaceuticals per day from drug wholesalers. They rarely keep any backup inventory. Our discussions with representatives from the pharmaceutical wholesalers—the critical middlemen in the drug pipeline—indicate that they typically keep a fourteen-day supply on hand based on historical volume purchases need per service area.

In other words, not much! Pharmaceutical manufacturers typically have between a forty-five- and ninety-day supply for what they call "usual usage patterns." This does not mean they have a forty-five-day supply on the shelf ready for shipping, but rather have in-house and available the material needed to create a "normal forty-five-day" excess, if necessary.

Pharmaceutical executives universally agree that under existing circumstances they would be hard-pressed to provide large amounts of antibiotics needed for a large metropolitan bioterrorism event with the kind of speed that would be necessary to make a difference.

Getting health care providers in other regions to give up their precious supplies is unlikely—especially in a situation in which they might fear a local attack themselves. I interviewed a number of hospital pharmacists, who all agreed that they would urge their administrators not to release antibiotics in the event of a distant attack, out of fear that they might need those same supplies for a yet-unrecognized release in their own towns.

All of this helps explain why the federal government has recently built up a national pharmaceutical stockpile

for use in bioterrorism crises. The CDC's National Pharmaceutical Stockpile Branch currently has on hand six
"push packs"—prepackaged shipments of enough antibiotics for 86,000 people. CDC has contracts with manufacturers to be able to get doses for millions more within
36 hours, bringing total coverage to over 5 million people.
That's a good start, but it doesn't mean that the problem is
solved. First, in an attack in New York City or Los Angeles, the 5 million treatments will go quickly. Also, as in the
anthrax scenario, other areas of the country will witness a
surge of patients with similar "early symptoms," making it
necessary to consider multiple events. This will easily
stretch the 5 million treatments to exhaustion. That's when
the next level of panic kicks in.

In addition, the logistical problems of getting the medicines
to the people who need them will quickly overwhelm most
health departments. Having to scramble to get antibiotics
and vaccines to a large population isn't as rare as you might
think it is. In January 1995, our team at the Minnesota Department of Health responded to a meningitis outbreak in
the town of Mankato by vaccinating thirty thousand residents in four days. We also supplied two days' worth of
antibiotics, along with the one dose of vaccine. It occurred
under the watch of one of the best health departments in
the country and it stretched us to the very limits of our
ability. Now imagine needing to vaccinate the 2.5 million
people living in the Twin Cities metropolitan area.

   To understand how things can quickly escalate out of
control when an outbreak strikes, let's go back to Bartlett's

scenario. The Baltimore area does have a strong system for responding to disaster, Bartlett says: "There is one person or one group that is coordinating the events and one point of contact that initiates the relevant cascade of events necessary for a response."

Bartlett notes that the influenza epidemic going on at the time of his experiment had already taken a heavy toll on the hospital: it was on "blue alert," meaning that all twenty-eight emergency room beds were filled. The hospital was filled as well, as was just about every other hospital bed in the city of Baltimore. This, says Bartlett, is a sign of the times: hospitals these days run at near capacity all the time. "The pressures of managed care have resulted in a health care system that has minimal elasticity, so on February 13 there were no beds for an anthrax epidemic." There were only thirteen thousand usable hospital beds in the entire state, Bartlett says, with little capacity for a surge in cases of any kind. "A recent large fire in Baltimore demonstrated that the city could not handle 100 casualties."

As Bartlett showed, health care trends like managed care and cost containment have contributed to a loss of flexibility in hospitals and clinics. Each winter, communities throughout the United States experience a "hospital inpatient capacity situation," which is a bureaucratic way of saying that there aren't any beds open for new patients because so many beds are occupied by people with influenza. In January through March 1998 the Los Angeles news media chronicled the crisis in patient care in southern California due to routine influenza cases. Severe bed and staffing shortages were widespread. In a

January 4, 1998, *Los Angeles Times* article, Susan Abraham and Julie Marquis reported on how local hospitals had to revert to "disaster plans" just to meet the needs of an increased number of flu cases. At Holy Cross Medical Center in Mission Hills, officials said emergency room patients who need admission must wait eight or nine hours for a bed. Dr. Brian Johnson, director of the emergency department at White Memorial Medical Center in Boyle Heights, said, "In my hospital, we're setting records. It's very hard to get beds. In fact, when you get a surge like this, we don't have enough beds. We have not had enough beds in the hospital for six weeks."

The same story is often played out throughout the United States each winter. The influenza season of the winter of 2000 taxed hospitals and clinics to the breaking point as well, but public health officials repeatedly noted that the sheer number of patients was no worse than what could normally be expected in a flu season. The difference? Not enough beds and staff to meet "predictable needs." I felt a chill when I heard a senior CDC official recently remark at a national meeting that "if the routine level of seasonal influenza stresses the system that much, I can't begin to imagine what a pandemic situation [influenza] or a bioterrorism event will do. It's beyond comprehension."

Other public health and safety officials understand that. Ellen Gordon, director of the Iowa Division of Emergency Management, testifying before Congress on behalf of the National Emergency Management Association in September 1999, warned that no local officials really feel prepared to face the consequences of an attack that uses weapons of mass destruction. "As a whole, the state directors of emergency management believe that most state public health systems are unprepared to respond to a WMD incident," she said. Gordon explained that there is a wide range of capabilities at the local level, but warned that the national focus on America's biggest cities for equipment, personnel, and training has left the rest of the country unguarded. She also said that there is little coordination of information between the medical and law enforcement communities. Perhaps most important, she said, neither "public health services nor private hospitals are equipped to handle WMD issues related to decontamination, mass casualties, and mental health care for victims, first responders and the community at large."

Tara O'Toole, a senior fellow at the Johns Hopkins Center for Civilian Biodefense Studies, issued a similar warning at the September 1999 hearing. Pointing out that a bioterrorism attack would "require a response that is fundamentally different" from response to chemical or bomb attack, or even an earthquake or fire, she noted that "the speed and accuracy with which physicians and laboratories reach correct diagnoses and report their findings to public health authorities will directly affect the number of deaths, and—if the attack employs a contagious disease— the ability to contain the epidemic." Yet, she noted, "few, if any, practicing clinicians have ever seen a case of smallpox or anthrax or plague. Only a handful of laboratories have the ability to identify definitively the pathogens of greatest concern." In other words, the hospital labs can find what they usually find—*E. coli*, strep, and the usual bugs that burden mankind. Only specialized government labs have the training and resources to do more—and getting samples to them could add to the delay in finding out what the disease causing the crisis is.

A full-blown outbreak, O'Toole warned, would utterly overwhelm local health care systems. "Few, if any, recent disasters on American soil have resulted in large numbers of patients needing immediate and sustained medical care," she said.

Sounding a note similar to Bartlett's warning about Baltimore's overworked health care system, she said that "hospitals, which thus far are almost entirely absent from any bioterrorism response planning activities, are already overburdened. Few cities have sufficient numbers of un-occupied hospital beds, staff or equipment to absorb even a moderate influx of severely ill patients."

The best-case scenario of attack would have the same result, says Dr. Edward Eitzen of the U.S. Army Medical Research Institute of Infectious Diseases at Fort Detrick, Maryland. Eitzen performed a simulation exercise known in the jargon of the field as a tabletop incident, in preparation for the 1997 G-7 summit in Denver. It was based on a hypothetical terrorist event that seems almost rosy in its particulars but chilling in its outcome. He suggested a release of anthrax within a shopping mall with just 10,000 people present—a tenth the number that can be found on a weekend at larger malls, and a thirtieth of the population of the biggest malls during holiday shopping madness. Of those 10,000, he estimated that 9,000 would be exposed.

Eitzen even went so far as to imagine that the terrorist would call in a warning after twenty-four hours that he had released the deadly agent, allowing public health officials to set up a crash inoculation and antibiotic provision program that in one day would reach 90 percent of those affected. It doesn't get much rosier than that. Within those conditions, Eitzen suggested, there would be 4,950 people hospitalized, with 2,925 requiring treatment in intensive-care units. Ventilator therapy would be required for 2,601 patients, and 855 would die. Eitzen compared this crush of humanity needing treatment with the largest-ever deployment of U.S. military assets in history, during the Gulf War. At that time, the military had 13,000 hospital beds, and only 1,300 were in intensive-care units—half of what would be needed under Eitzen's very optimistic scenario.

Like Bartlett's frightening experiment in Baltimore, Eitzen's tabletop exercise could have served as a wake-up call for Americans who don't yet recognize the risks of biological terrorism. But it received scant press coverage. Most of the stories about the counterterrorism training program in Denver and several other U.S. cities stressed the good news—that the training was taking place—and not the bad news about the gravity of biological terrorism itself. The experts hear these facts over and over, and frighten each other. The general public, which needs to know the details and hard truths so that they can demand action, doesn't get the message.

Whole regions don't have the facilities to face even a moderate outbreak. In 1998 our group at the Minnesota

Department of Health surveyed the medical resources for the entire state and found that of its 144 hospitals that provide acute care, only 60 had one of the most important tools for taking care of patients with highly contagious diseases: negative-air-pressure rooms.

Crucial for truly isolating someone with a highly communicable disease like smallpox, a negative-pressure room is equipped with machinery that ensures that air and the bugs floating through it do not leak out into the rest of the hospital. Without it, infection can spread quickly, and with disastrous results.

And 60 units are not much. Altogether, those 60 negative-air-pressure units in Minnesota's hospitals served just 465 beds—some of them had only 1 such bed, some as many as 75, but the mean number was 8. And only 108 of the beds were within the critical-care units that would be needed for the treatment of a dire illness like smallpox. Minnesota didn't even have enough body bags on hand to handle a limited number of deaths; it had only enough to cover the crash of a large jetliner at the Minneapolis–St. Paul International Airport.

Laboratories would also be inundated and unable to perform their critical functions. In a smallpox epidemic, one of the most important early issues to resolve is determining whether new patients actually have the disease or have come down with something less threatening. If smallpox cases were being sent to mass treatment centers, the right diagnosis could mean the difference between life and death; patients who were not infected when they were sent to such a center almost surely would be once they got there. Ruling out chickenpox and flu would be critical—but very few laboratories are equipped to perform the kind of testing necessary to detect smallpox or other agents likely to be used in biologic terrorism. Unfortunately, getting those hospitals to plan ahead presents just as daunting a challenge: the leaders of the hospitals aren't interested in joining government-sponsored exercises. They are "preoccupied with a welter of urgent issues associated with the changing and financially competitive terrain of modern health care," O'Toole says. "Most hospitals are not in a position to accept unfunded mandates, and are unlikely to respond to bioterrorism response plans unless the nation establishes a thoughtful menu of incentives and programs that enable them to do so." The cutthroat competition between health care systems prevents cooperation as well.

The other issue that would almost certainly spell disaster at any hospital confronted with an outbreak is panic. Many health care providers would clear out the moment that a case of smallpox was even rumored.

One thing is certain: if smallpox were to hit a modern American hospital, it would prove just as contagious as it has through the ages. That was proved as late as 1970, in an outbreak in West Germany. A West German electrician returning from Pakistan was hospitalized on January 11 of that year with high fever and diarrhea. Doctors initially decided that the man had typhoid fever and isolated him in a private room on the ground floor of the hospital; he had contact with only two nurses over the next three days. The rash developed on January 14; the diagnosis of smallpox was confirmed by the 16th, at which point he was taken to a special isolation hospital, and more than 100,000 people were vaccinated.

As it was, however, nineteen cases occurred in the hospital. Three nurses contracted the disease; one of them died. The disease traveled far: to eight people on the floor above the patient's and nine on the hospital's third floor. One of the victims had passed through the hospital in less than minutes and never got closer than thirty feet from the patient's door. Luckily, the hospital had already been closed to visitors because of an outbreak of influenza; otherwise many more people might have been infected; it's important to remember, too, that vaccinations against smallpox were common at the time.

In the 1972 outbreak in Yugoslavia, things went far more badly. The first cases were correctly diagnosed four weeks after the first patient became ill, but by then, 150 persons were already infected; of these, 38 (including 2 physicians, 2 nurses, and 4 other hospital staff) were infected by a young teacher. The cases occurred in widely separated areas of the country—and among a population that had reasonably current smallpox vaccinations. By the time of diagnosis, the 150 secondary cases had already begun to expose yet another generation, and, inevitably, questions arose as to how many other yet-undetected cases there might be. Chaos and panic resulted. The nation launched a mass vaccination campaign, and checkpoints along roads were established to examine vaccination certificates. Twenty million people received vaccinations. Hotels and residential apartments were taken over, cordoned off by the military, and all known contacts of cases were forced into these centers, under military guard. Some

10,000 persons spent two weeks or more in isolation. Meanwhile, neighboring countries closed their borders. Nine weeks after the first patient became ill, the outbreak stopped. In all, 175 patients contracted smallpox, and 35 died.

If smallpox were to return today, the situation could be far worse still, because the large stockpiles of vaccine that were handy during the 1970s outbreaks are no longer available. With smallpox supposedly vanquished, no one produces the vaccine anymore. The CDC estimates there are 60 million doses of the vaccine worldwide, with 15.4 million doses in the United States—far fewer than the 40 million doses its own experts say it needs.

The federal government has spent millions to kick-start the production of new vaccine from a private company, but the fruits of that effort are still years away. More than four years ago, the Department of Defense began looking into creating a new smallpox vaccine and building a stockpile for the military, but the effort is so enmeshed in politics and grant jockeying that one military officer with direct experience with the program described it to author Richard Preston as "a fucking disaster." The cost of the 300,000 doses of a new smallpox vaccine is $22.4 million: that completely inadequate stockpile won't be ready until the year 2006 at the earliest.

The Department of Health and Human Services has also begun looking into the need to create a civilian stockpile of 40 million doses, Preston reports. The CEO of the company with which the military contracted told Preston that accomplishing that task would be far more expensive and complicated than supplying vaccine supplies to the military.

Because of the variation in the civilian population, he said it would be necessary to factor in the vaccination risks for the very young and the very old, those with compromised immune systems from cancer treatments and AIDS, and more. "It's not just scaling up the manufacturing." Retired army general Philip Russell, M.D., who dealt with the Ebola Reston outbreak in 1989, told Preston that he and many others worry that the civilian effort "won't deliver the goods without wasting an inordinate amount of money."

Worst of all, there is concern that the existing stockpile and the tools necessary to treat people with it have degraded to the point that much of the supply might not be usable. The freeze-dried crystals of vaccine are derived from the cattle disease vaccinia and are stored in vacuum-sealed tubes with rubber stoppers. CDC investigators found condensation in many of the tubes, which might indicate that the stoppers had decayed; that, in turn, could hinder the effectiveness of the vaccine for several million doses. The fluid used to dilute the vaccine had also decayed, so new diluent needs to be manufactured. Even the special two-pronged needles necessary to scratch the vaccine into the skin are in short supply, with fewer than a million left in the world—and no one manufactures them anymore.

As if all that were not troubling enough, the drug used to treat patients who react badly to the vaccine, vaccinia immune globulin (VIG), is also in short supply, which will hinder any attempt to vaccinate large numbers of people. There are only enough doses to treat 675 cases—people who might otherwise die from the vaccination—and even that stockpile can't be used until a troubling discoloration that has affected many doses is explained. In other words, we are years away from being remotely ready for the specter of smallpox.

7. THINGS FALL APART

As the twin epidemics—smallpox and fear—spread, people have stopped going out. The streets, at midweek, are eerily silent. Barbara Bradburn's commute time would be slashed if she were commuting; instead, she's living at the hospital in a sleepless haze of work and unending crisis.

When she retreats from the terrors of the wards to her office to lie on her couch, she finds that her nerves are too jangled and raw to allow sleep. Sighing, she switches on the small TV on her desk; being able to watch it at all is a luxury, thanks to the hospital's backup generator. Some Chicagoland neighborhoods no longer have power, while in others water and sewer lines that broke in the first cold snap go unattended. No one is coming in to work to keep the systems running. The el doesn't roar through the city. Chicago has the feel of a ghost town.

Only one local channel is still operating. Bradburn surfs the remote, but there's no escaping the epidemic; it's wall-to-wall, 24-7 coverage everywhere.

"Yes, Peter, the Chicago crisis is being felt throughout the nation," a reporter says in a live report from Wall Street. "With the city of broad shoulders' critical commodities exchanges stilled, America's financial and commodities markets are in turmoil, disrupting services nationwide."

The screen now shows a view of the panicked traders on the floor of the New York Stock Exchange. "Stock and commodities markets have crashed, the trillions of dollars in financial gains of the long stock market boom wiped out."

The image cuts again, this time to a view of the empty Chicago stockyards. "Beef sales have plummeted across the country because of rumors that smallpox has contaminated Chicago's meat supply."

The haggard face of a spokesman for the beef industry fills the screen. "There is nothing to fear from beef—the American beef supply is the safest in the world, and we're making sure that Chicago's beef is not on the market," he says. "But it's hard to counter the rumors; truth is no defense in a panic."

With staffing at Cook County Hospital down 60 percent, Bradburn is one of the most senior medical officials left at the hospital and she takes command of the smallpox crisis. But she's making it up as she goes along; the city never prepared for anything like this. Only a limited amount of vaccine has made its way to Chicago so far, and it was given to health care workers to keep the health care system going—despite inflammatory cries of special privileges for the doctors. Even then, Bradburn has no idea whether she got the vaccine in time to ward off the disease. CDC officials tell her that getting the vaccine even a few days before exposure would likely have protected her, but no one is sure what it means to get it a few days after exposure.

The news anchor turns coverage over to a live press conference. The President, looking exhausted, announces that there are now thousands of cases of smallpox in midwestern cities, with the greatest concentration in Chicago. "As laid out in Presidential Decision Directive 39, FBI Director P. William Simmons is in charge of the situation on the ground," the President announces.

"Wonderful," Bradburn thinks to herself. "Now somebody's in charge. But he's there, and we're here."

The media fall into the familiar din of shouted questions—who is at risk, who is responsible, are there any leads—until one voice emerges from the chaos.

"Mr. President, the FBI says it still has no leads, and that no one is taking credit for the attack. Do you have any further information for us on that?"

"Mr. President! The CDC says there are fewer than 16 million doses of vaccine, maybe even only 7 million doses, and 6 million people in Chicago alone. With millions of people in the other cities needing to be vaccinated, the stockpile can't possibly do the job. Why don't we have more vaccine after all this time?"

The President blinks at the question, stares at the reporter for a moment. He signals to his chief spokesman, who shouts, "No further questions!" as the President turns and quickly leaves the room, away from the cameras and the lights.

The phone rings; Bradburn picks it up. The person on the other end is frantic, yelling into the phone. "I'm sorry," Bradburn says, holding the receiver away from her ear. "We don't have enough vaccine to distribute it broadly yet, and even when we do get it we're going to have to give it to the people treating patients, working the ambulances, the police, the firefighters. There's nothing more I can do."

"I'll come down there and kick your ass!" the man on the other end of the phone shouts. "I'll kill you!"

"Too late, asshole," she replies. "I'm already dead." She hangs up.

On a highway leading out of town, shoe salesman Bill Reynolds sits in his car at a roadblock. He knows by now that he has been exposed to smallpox—one of his coworkers even came down with the virus and died. But when he developed a fever, he had refused to be quarantined, even when confronted by the officials who had checked work records from the mall and were making the rounds of employees, enforcing the crisis response plan.

"Officer," he had said, "I might have been at work the day the disease got spread around, but I don't think this is smallpox—hell, I don't even feel that bad! As for Joe, at the shop, well, he stopped coming to work well before his rash had appeared. I've done my homework—Joe hadn't been contagious yet."

"Mr. Reynolds, I don't make these rules—and nobody can be sure that you're not still infected."

Reynolds's voice had risen then. "You know that entering that 'treatment facility' of yours would just be a death sentence," he responded. "If I wasn't exposed before, I will be as soon as I get there." Reynolds had heard, too, that no one was actually getting any treatment. "Those aren't hospitals—they're concentration camps!"

He had told them he was going into his house to get a suitcase, but sneaked out the back way and raced off in his car.

Now he's just as trapped as if he'd stayed. The line of cars behind him is growing longer, pinning him in place. He turns on the radio, looking for an oasis of calm. Most of the area stations have stopped broadcasting, but the nostalgia station he likes is still on the air, playing Christmas music. Reynolds is briefly reassured; it's the Christmas season, they're playing Christmas music right on schedule. "Bing croons eternal," he says.

He looks at the cars ahead and realizes that people are being turned back: no one is allowed to leave the city until the epidemic is under control. His calm shattered, his fingers grip the steering wheel as the guards move down the line toward his car. Reynolds guns the engine and pulls out of line, hoping to run the roadblock; the lawmen draw their guns and fire.

Taking enough bullets to speed the process of dying considerably, he slumps at the wheel. He fades into unconsciousness to the sound of Bing Crosby wishing him a white Christmas.

WHEN FRANKLIN DELANO Roosevelt in his first inaugural address told a nation shattered by economic depression that "the only thing we have to fear is fear itself," he knew the awful damage that panic could wreak on a society. "Fear itself" is a formidable foe, and one of the most important to confront in the event of a bioterror outbreak. Margaret Hamburg, assistant secretary for Planning and Evaluation at the U.S. Department of Health and Human Services, recognized the importance of the problem in March 1999 congressional testimony:

> Terrorist attacks are intended to create some combination of illness, injury, suffering, death and economic loss—all of which increase the likelihood of behavioral, psychological and social disorder. Bioterrorism, with its implication of death arriving imperceptibly through the air we breathe, poses a new constellation of threats to the resilient human mind and to the power of both large and small communities to survive intact. In addition to the potentially massive numbers of physical casualties or deaths, bioterrorism, by threat or in fact, will create a devastating number of psychological casualties. A bioterrorist event is different from all other forms of terrorism in its potential to precipitate mass behavior responses such as panic, civil disorder and pandemonium. This is especially true if the bioweapon used is a communicable agent that spreads disease in successive waves of transmission. We could expect a bioterrorist attack to seriously disrupt local and regional economic functioning over many weeks or months since the "damage" that is inflicted is not to material infrastructure but to the human infrastructure—a kind of damage that takes considerably longer to repair.

It's that "human infrastructure" that this chapter deals with. Like spots of rust causing the collapse of a once-mighty bridge, fear corrodes the structure of society. What is it, then, that holds a society together in harsh times? What factors strengthen the social structure against chaos?

One scholar who has studied the delicate balance between rights of individuals and needs of the community is Amitai Etzioni, a professor at George Washington University and founder of the Communitarian Network. Etzioni believes the result of an attack would be extreme: "In the case of a really serious attack of biological terrorism, we would suspend the Constitution. . . . All bets are off," he says. "I'm not advocating, I'm just reporting."

Whether people would accept that situation, Etzioni argues, rests on two conditions. First, leaders must convincingly explain to the public that there is a "genuine threat"; then leaders have to "thoroughly explain what you're doing and why you're doing it." Most smokers accept restrictions such as public smoking bans, Etzioni says, because they have come to believe that smoking harms others and understand why the bans are being imposed. But because the nation's leaders never adequately demonstrated the threat of alcohol or the logic behind restricting access to it, Prohibition was a flop. "When you don't do those things, you lose," Etzioni says.

That's the way, of course, that it works in popular fiction. In such thrillers as *Outbreak* and the Tom Clancy novel *Executive Orders*, the President simply does what has to be done, and the soldiers show up. In the Clancy novel, the President does order a nationwide shutdown to stop the spread of a terrorist Ebola outbreak. The President's advisers warn him that the actions would be unconstitutional—one asks, "If we flaunt our own laws, then what are we?" President Jack Ryan responds, "Alive, maybe."

Clearly, fear would sweep the nation after any bioterror attack; keep in mind, though, that the highly contagious smallpox release described in this scenario would cause unprecedented rates of death from a single act of violence.

Recent history is a poor guide to the scope of the disaster. The notion that each wave of smallpox infection will be ten- to twentyfold the size of the one before it, for example, is based on research done during the days when much of the world's population was getting regular vaccinations and there was a fair amount of natural resistance. Today's population has little or no such resistance: the disease would spread like fire through dry tinder. Since the nation's vaccine stockpiles are inadequate to help quench even a regional outbreak, there is virtually no protection from the microscopic onslaught. Containing the damage would take months.

In 1997 the CDC evaluated the economic impact of a bioterrorist attack for each of three different biological agents: anthrax, brucellosis, and tularemia. Their model showed that the expected economic impact from such an attack would range from $477.7 million to $26.2 billion per 100,000 persons exposed. The authors of the study concluded, "These are minimum estimates."

So, too, the disruption laid out in the preceding scenario could be a conservative version of the results of a successful bioterror attack. The interlinked financial, transportation, utilities, and communications networks form a national infrastructure that, for all its size and robustness, is potentially very fragile. In 1997 the President's Commission on Critical Infrastructure Protection put it this way in discussing the destabilizing results of possible acts of cyberterrorism by hackers: "Our security, economy, way of life, and perhaps even survival, are now dependent on the interrelated trio of electrical energy, communications and computers." The kind of hit that a bioterror attack would have on a city like Chicago would have direct—and catastrophic—effects on all three systems nationwide.

Disruption wouldn't stop at borders, either: wherever fearsome diseases rear their head, the international community reacts with horror—and tries to keep the damage from spreading into their lands. The 1995 outbreak of Ebola in Kikwit, Zaire, shut down much of the travel and commerce from the West to Zaire—and, indeed, to other African countries. During the West German smallpox outbreak in 1970, service station attendants reportedly refused to even serve drivers whose cars bore the license plates from the affected area.

In such times, government has to respond with a calm voice and a firm hand; it takes the manner of a Roosevelt to ease the tension and get things moving again, and it takes a government that knows what it's doing to respond effectively.

A prominent legal scholar who has examined the issue says that the nation's laws don't lay out a clear path for that response—or even give government the authority to do so. Terry P. O'Brien, a former assistant attorney general for the state of Minnesota, has performed an in-depth analysis of the legal framework that would have to come into play in the case of catastrophic terrorism.

I've known Terry for nearly twenty-five years; he was the lawyer assigned to help me and my staff to deal with the jurisprudential tangle that comes with my department's work. In 1995 I asked him to examine what I saw as the weak structure to balance laws and rights in case of an attack, and to determine whether we could even legally enforce something as simple and necessary as a mass quarantine. At first he shrugged the inquiry off, saying that he'd be duplicating the effort of others: "You've got to be wrong," he told me. "Somebody has to be looking at this." To his shock, he found that no one had ever really addressed the patchwork of laws and regulations, and he began to dig deep. Tenacious and scholarly, he is now recognized as the nation's leading authority on the constitutional and legal problems that would accompany a bioterrorist attack.

He notes that at the end of the nineteenth century, the American people had a better understanding of the destructive power of epidemics, and so laws regarding control of epidemics were extremely tough. When a placard reading SMALLPOX or POLIO appeared on the door, the people inside knew they had to remain inside until the disease had run its course.

But two major shifts in society might make any new outbreak much harder to control. The first is the medical revolution in antibiotics and antiviral drugs, which have helped Americans to forget just how destructive a disease outbreak can be if not brought under control quickly. Even AIDS, for all of its destructiveness, is not very contagious compared to such diseases as smallpox; though AIDS patients have had to modify their lifestyles, there has not been a need for the kind of quarantine that a major outbreak of contagious disease would entail.

The second major shift is a revolution in civil rights—an explosion in rights of individuals beginning in the 1960s and 1970s that the nation is still adjusting to today. A broad attempt to redress violations of civil rights in the criminal justice and public health systems, the rights explosion set new standards for due process for anyone whose property is taken or whose liberty is curtailed—in itself, another of the great achievements of twentieth-century democracy. But this expanded respect for individual rights has changed the landscape for situations in which a broad quarantine decree might be necessary, O'Brien says. "If somebody tried to enforce a quarantine today, the response might be 'screw you,' " he muses.

O'Brien says that any legal response to bioterrorism would have to recognize how the world has changed and attempt to strike a new balance between the rights of the individual and the needs of society—and, he insists, it should be done now, before such laws are needed. Having researched federal laws on the books, he has seen nothing that would remotely address keeping order during a bioterrorism event. "We have to deal with this now, in a calm way," he says. The alternative—decisions made during a crisis—is likely to have the hallmarks of overreaction. Civil rights could easily be trampled, he warns, just as the rights of Japanese Americans were trampled in the mass internment during World War II.

Nowhere has the tension between the rights of individuals

and the medical needs of society been more clear than in the AIDS epidemic. Early response to the epidemic included fear-based proposals, including mass quarantine of infected patients, that would have made a mockery of the Constitution, O'Brien says. After all, HIV isn't transmitted via the air we breathe. But staunch libertarian sentiments within the AIDS community also stopped or delayed legitimate public health initiatives that have been employed in outbreaks of other diseases and that could have helped contain the spread of AIDS, such as disease surveillance systems that could count infected patients.

Some states are trying to strike a new balance. Minnesota, for example, has developed laws that respect individual rights and the needs of society. The state does have HIV reporting by name, and a separate program for addressing the spread of sexually transmitted diseases by controlling the activities of prostitutes and people known to have the diseases but remain sexually active with unsuspecting partners. That law is carefully crafted to meet the goal of containing the spread of sexually transmitted diseases, however, and allows the infected person three warnings before even being taken before a judge. The balance, O'Brien argues, is perfectly consistent with the needs of society in the case of sexually transmitted diseases, but would be utterly inadequate in the face of a rampant outbreak of a respiratory-transmitted deadly disease.

New York has created tough quarantine enforcement programs to meet the dire threat of antibiotic-resistant tuberculosis in a way that tips the scale away from individual rights and toward the needs of the community. Today a patient diagnosed with tuberculosis will be taken to a special facility and held there for months until the course of heavy-duty antibiotics is fully run—the only way to ensure against a new generation of bugs that can resist even the remaining effective antibiotics.

O'Brien worries that the laws that currently purport to address these issues of keeping order after a terrorist attack have not yet addressed these ground-level issues of respecting rights while taking decisive action.

Despite this lack of clarity, there's little doubt that, as Etzioni suggests, a real-life bioterrorist event would trigger drastic action by government. When the Supreme Court decided to affirm the herding of Japanese Americans into camps during World War II, even dissenter Justice Owen J. Roberts noted that the military would have authority under presidential direction to remove people "from the area where a pestilence has broken out." No special legal measures were taken to combat the most compelling public health crisis of the twentieth century— the 1918 influenza epidemic that killed more than half a million Americans—but Roberts clearly reasoned that epidemics can change the rules.

History shows us that the result has not been good— not for the Constitution and not for the people governed by it. The laws that lay out the federal government's authority in time of crisis don't give much guidance. Even more troubling, O'Brien and others who have looked at the laws currently governing response to terrorist attacks say those laws are poorly designed. The flaws in the laws might lead to rights violations on the one hand and legal challenges on the other that could prevent government from being able to do even those legitimate things it has to do in a crisis. Consequently, what may appear as arcane policy issues are in fact vital questions that could have a dramatic impact on all of our lives.

The key legal document describing the government response to a terrorist event is Presidential Decision Directive 39 (PDD-39), issued in June 1995. A response to the Oklahoma City bombing and the Aum Shinrikyo attacks, its full text is classified. Unclassified versions of the document are available, however, and O'Brien was able to glean more information in interviews with officials familiar with the classified portions.

The directive divides response into two areas: crisis response and consequence management. The first term describes methods of blocking terrorist acts before they can happen and is coordinated overseas by the Department of State's Office of Counterterrorism and domestically by the FBI. Consequence management concerns the response to an actual attack.

PDD-39, however, is riddled with problems. O'Brien cites criticism by a Pentagon insider who served on a Chemical-Biological Incident Response Force during the Atlanta Olympics. He concludes that PDD-39 lays out artificially neat lines between crisis management and consequence management, separating jurisdiction and command in a situation where no such bright lines will actually exist. "In a WMD situation, domestic or international, consequence management is the crisis," the critique states.

Other authors who have looked closely at PDD-39 have found similar flaws. Richard A. Falkenrath and co-authors suggest that the presidential directive is a vague document that provides no blueprint for how government should improve its ability to meet a WMD crisis. They argue that it places too much responsibility for coordinating response to a terrorist event on the shoulders of the FBI, which has too little experience successfully coordinating large-scale, multiagency operations, and fails to delineate the various responsibilities of federal, state, and local government.

Another of the laws that govern response to terrorism is the Nunn-Lugar-Domenici Amendment to the National Defense Authorization Act for fiscal year 1997. Also designed to enhance preparedness for attack, the law's first-year appropriation included $50 million to the Department of Defense for training first responders, creating medical "strike teams" and domestic terrorism rapid-response teams, and planning and conducting training exercises. But the program never received the full support of the Clinton administration, and it focuses more on responding to traditional terrorist violence, emphasizing the roles of soldiers and police rather than doctors, nurses, and public health workers.

There are a number of laws that the government can invoke in times of crisis, but the President's authority to use those laws is not clear, and the relevant case law is conflicting. Title 10 of the U.S. Code, for example, allows the President to use emergency powers. This usually occurs after the request of a governor and empowers the President to send in federal troops to control civil disorder. Invoking this presidential power to mobilize the Department of Defense to respond to a bioterrorism crisis is envisioned in the Nunn-Lugar-Domenici bill.

Those powers, however, face a sharp limiting factor: the 1878 Posse Comitatus Act, passed by Congress during the period following the Civil War to restrict the government's ability to use the military to keep civil order. That law, more than a hundred years old, has had remarkable staying power. A number of court decisions surrounding the use of federal troops in the American Indian uprising around Wounded Knee, South Dakota, in 1973 found that the military should not have been used (though the case law is inconsistent on this point).

Attorney General Janet Reno testified in 1998 that the Posse Comitatus Act would indeed restrict the military from executing laws in the case of bioterrorism. A 1984 disaster-relief law, however, allows the President to send troops in to deal with a national disaster. That law includes this definition of emergency: "Any occasion or instance for which, in the determination of the President, federal assistance is needed to supplement state and local efforts in capabilities to save lives and to protect property and public health and safety, or to lessen or avert the threat of catastrophe in any part of the United States." But the law also restricts the ways in which active-duty soldiers can be deployed, mandating that they operate under the direction of the Federal Emergency Management Agency, and prohibiting them from detaining, arresting, or otherwise engaging civilians in other forms of restraint, roles that they could potentially be called upon to play in a bioterror situation.

An example of the kind of conflict that could emerge occurred in 1992, when President George Bush, at the request of Governor Pete Wilson of California, sent U.S. Army and Marine forces to help quell the riots that followed the Rodney King verdict. Once the 3,500 soldiers were in place, and the military authorities held their first meeting with the L.A. police chief and county sheriff, "it became clear that no one had a clear perception of the proper role of military forces in an emergency," O'Brien writes. "The Chief wanted to partition the city into military districts. The Sheriff, in a 'rent a soldier' fashion, wanted to have various military forces allocated to police units. The Joint Force Commander, believing he was constrained by provisions of the Posse Comitatus Act, did not believe he could legally participate in law enforcement activities."

Scholar Thomas R. Lujan points out that soldiers are not trained as law enforcement officers and are not accustomed to abiding by the legal boundaries placed on law enforcement officers. "Before decisionmakers bring our military forces to bear," he writes, "the situation must be so potentially harmful . . . that the United States must react as if it is an act of war—not just a crime. In domestic terrorism requiring a military response, the armed forces are not just 'adjunct police,' they are military forces operating under military rules of engagement."

If an attack can be determined to have come from a rogue state, the President could also invoke general executive powers under a number of laws, such as the War Powers Act.

O'Brien concludes that whatever the legal status of a federal takeover during a crisis, the stakes for the Constitution are exceedingly high. "Even under the patina of some colorable legal authority, imposition of martial law even in a severe outbreak would be a first time event in the history of the country and would severely impact the liberties not only of our citizens but the integrity and fabric of our government."

The alternative to those draconian measures isn't very promising either, however.

Dealing with a bioterrorism event will be the toughest job a cop ever has to face. I have spoken at conferences with police and sheriff's deputies present and asked them point-blank, "What are you going to do with a patient who has smallpox—or even who has been exposed and is not yet ill—and refuses to be quarantined? Somebody who runs away from the hospital? Somebody who has become a human weapon, capable of spreading the infection to hundreds or even hundreds of thousands of people? Will you be prepared to stop them when your public health authorities determine they pose a major risk to the community? To shoot to kill, if necessary?" At a talk I gave in 1998, not one of a large group of deputies said he would do it. Not one of them said he'd even go near enough to restrain or transport such a patient.

O'Brien has repeatedly presented his scholarship and concerns to officials at the federal level—most recently at a national terrorism response conference in Pentagon City, Virginia, in February 1999. The response, he says, is always the same: there is no response.

Sometimes we can't even get the response to hoaxes right. Even though they reflect little of the real risk of a bioterrorism event, hoaxes demonstrate the disconnect between most law enforcement and first responders on one side and public health and medical experts on the other. The wake-up call for Americans might have come in April

1997 when the Washington, D.C., offices of the international Jewish organization B'nai B'rith received a suspicious package. It contained a leaky petri dish and a note that indicated that the package might contain microorganisms causing plague or anthrax. As fear gripped those in the building, emergency workers created a crisis of their own, with humiliating outdoor strip-decontamination procedures for those potentially exposed and harsh conflicts between police and emergency medical responders, who saw the situation differently. The emergency medical team tried to treat the incident like a chemical or disease outbreak incident, setting up "hot" zones to control access and keep infection from spreading. Police treated the incident like a bomb scare and repeatedly crossed from zone to zone in pursuit of their investigation. While the victims sweated, the responders—who have a history of tense relations—squabbled; one police officer even punched an emergency medical worker who insisted that the police take off their uniforms and shower. If nothing else, it showed that such incidents require training, clear procedures, and quick action.

One of the fundamental tensions that will emerge within any bioterrorism event already comes up whenever crime and medicine mix: cops and docs see the world differently. When they come upon a disaster scene at the same time, they have separate and conflicting goals. The police immediately begin their investigation and need to preserve the crime scene as much as possible in order to find the clues that can be obliterated in a rescue effort. The medical teams, on the other hand, are trying to treat victims and don't care what they tear up in the process. They aren't trained in preserving a crime scene and have no interest in doing so. For their part, many police don't want to be bothered by the necessities of medical rescue efforts and prevention of contamination.

As I write this more than two years after the B'nai B'rith scare, some attempt has been made to address these problems—but there is still much to be done. Even during the first weeks of January 2000, when anthrax hoax letters were again delivered to Planned Parenthood clinics around the United States, local police and fire departments grossly mishandled the situations. And public health? They weren't even involved. Unfortunately, much of what *is* being done bears the same flaws that caused the problems in the first place.

The problem is that so much of the planning focuses on the needs of law enforcement that the public health and medical experts are often pushed to the side. I was asked to take part in planning the Justice Department's major national exercise for responding to domestic terrorism, known as TOPOFF. In May 1999 I was sitting in a conference center in Chantilly, Virginia, where 150 people had been brought together by the Office of Justice Programs' Office for State and Local Domestic Preparedness Support (OSLDPS, but who would ever use an abbreviation like that?). As they went on and on about weapons of mass destruction and how police and firefighters would be "the team" to respond to their use, it became increasingly clear to me that they had no clue. The other members of my small breakout group were going along with the exercise as if firefighters and cops were going to be able to take on an outbreak of anthrax or smallpox. "You can do this," I told Washington, D.C.'s deputy chief of police, "but it's just not relevant." I looked around the room—and realized that of the people there, I was one of only *three* with a background in public health or medicine. Two of us were there almost by accident. That's the problem in a nutshell: the federal, state, and even local governments just don't get it.

8. VIDEO GAMES AND TRICORDERS

Barbara Bradburn wonders if she's hallucinating. She shakes the cell phone—the landlines stopped working the day before—to make sure that it's actually working. "Pete, tell me it's a joke," she begs. "And then remember that I'm not in the mood for jokes."

"No, I'm serious, Barb. They're fighting over who owns it— people are dying, and these jerks are in a goddamned **pissing** match."

The man on the other side of the phone, Pete Martin, was a professor of Bradburn's in medical school; he had worked in the

1960s on the WHO team that had knocked out smallpox a generation before. So when the crisis hit, he had been called to the White House as chief medical adviser.

"You wouldn't believe it, Barb—there was the President of the United States looking around the cabinet room at law enforcement guys, public health, intelligence, defense, you name it. He wants progress. He wants action. He wants to make sure he doesn't get hit with the blame for this. And the bureaucrats are talking about who 'owns' the case. FBI is swaggering around citing the national antiterrorism laws that put them in command of response, but oh, no, say the spooks from CIA, there might be a rogue nation involved, so CIA insists on staking its claim. A pox on all their houses, I say."

"Did they ever talk about us, Pete?"

"Well, they did finally let me do my dog and pony show. I tried to explain to them that full quarantine is the only viable option—and that something like martial law is the only way to make it work."

"I remember your damned lecture, Pete," Bradburn said, and then dropped her voice into an imitation of his: " 'Smallpox is like a fire—the quicker a fire wall can be built, the better.' I never thought I'd see it put into practice here, though."

Earlier in the day, Martin had also explained the procedure by phone to the police and volunteers who would have to create and enforce a quarantine program: quarantine teams have to go from door to door looking for people with symptoms. Anyone who shows symptoms has to tell whom they've been in contact with over the previous seven days, and then all of those people have to be tracked down, forcibly quarantined, and vaccinated. "It's a lot of work, but otherwise the disease spreads out of control," he explained. "It's a new Chicago fire, but this time the fire will cripple the nation if we don't act now."

Since the vaccine will take two weeks to build immunity—and the supplies are too limited to take on full-scale vaccination throughout the region—many of the emergency and medical workers will have to be outfitted with HEPA filter masks. For most, though, the masks offer little comfort or sense of security, and besides, they're in short supply.

As Barbara listens to her old friend, her eye wanders to the small television and CNN's coverage of her nightmare. There's a mob scene on the screen: people have surrounded a truck passing through a small town and dragged the driver out. An impossibly young local reporter is squinting in the bright sunlight and yells over the background noise, into the camera, "Bernie, the people here apparently thought the truck had come from the disease center in Atlanta and was carrying vaccine."

"Pete, what did they say about vaccine?" Barbara asks.

"That's the worst part, Barb. The worst. This guy from HHS goes into major-league hand-waving under the President's questions about the vaccine. There are delays, the stockpile is still in the pilot-project stage, and we can't try diluting what we've got until we have scientific proof it'll work. . . . You just can't bet on getting any more of the stuff, probably not until it's all over. Do the best you can, kid."

She closes the cell phone and puts it down on her desk, and lays her head on her arms. She cries herself to sleep.

The once improbable has become the inevitable. Are we prepared? By most accounts, the answer is no.

Despite significant efforts to combat terrorism and
improve national readiness, medical response
capabilities are not yet well developed or well
integrated into consequence management plans.
Providers are not trained to diagnose or treat
the uncommon symptoms and diseases of
unconventional warfare. Public health surveillance
systems are not sensitive enough to detect the
early signs of a terrorist-induced outbreak.
Hospitals and clinics lack the space, equipment
and medicines to treat the victims of weapons of
mass destruction.

—CONGRESSMAN CHRISTOPHER SHAYS

IT WAS OCTOBER 1999, and ABC's *Nightline* had just
finished a week's worth of extraordinary programs devoted to a single topic: biological terrorism. The program
played out a scenario of an anthrax attack in a city's
crowded subway system, with a death toll climbing to fifty
thousand. In an on-screen interview, the mayor of Atlanta
acknowledged that the program was an eye-opener for
him; he hadn't had any idea of the implications of a bioterrorism event or of how unprepared Atlanta was to respond. The scenario itself was appropriately chilling, but
what followed was *really* frightening.

On October 12 three of the top officials in the Clinton
administration appeared on the program to defend their
programs: Defense Secretary William Cohen, Health and
Human Services Secretary Donna Shalala, and Richard
Clarke, the national coordinator for security, infrastructure
protection, and counterterrorism at the National Security

Council. Under tough questioning from *Nightline*'s Ted Koppel, the three officials squirmed, blustered, and bragged about all that was being done. But to viewers with any familiarity with the programs they discussed, it was painfully clear that they were ignorant of the actual facts, misinformed, or worse:

- Clarke and Shalala spoke of the national stockpile of vaccines and antibiotics as if it were already in the process of being built up. While we've made progress with the antibiotic stockpile, the supply of smallpox and anthrax vaccines remain a critical problem.
- Clarke bragged of meetings in 157 municipalities and major preparedness exercises conducted with 26 cities, all but ignoring Koppel's point that the mayor of Atlanta the week before had admitted that he had heard nothing about such briefings. Clarke defended the exercises as valuable and said, "The local areas have to be able to hold until the federal response units arrive"—as if the doctors, nurses, and hospitals could somehow tell a disease to stop growing until the federal cavalry appears.
- Cohen, who two years before had brought into the same studio a five-pound bag of sugar that he said represented enough anthrax to wipe out a city the size of Washington, was asked if he had said at that time, "It's not a question of if, it's a question of when." He replied, "I could have."

- "How ready are we on a scale of one to ten?"
Koppel asked. Shalala said, "We're not ready
enough, but we're getting there, and we have a
strategy. We have a national strategy." Koppel
tried to press Shalala on the point. "I knew you
weren't going to give me a straightforward
answer," he said before shifting gears and stating,
"The fact of the matter is, you can't even tell the
country right now how long you think it will be
before you are going to be as ready as you can
get." Shalala responded obliquely with a
discussion of preexisting programs such as a
system for tracking infectious disease outbreaks
and research at the National Institutes of Health,
"so we're not starting from scratch."

The show was, in fact, a metaphor for Washington's approach to combating biological terrorism: a lot of talk with too little forward movement.

A senior federal official close to the bioterrorism preparedness activities called me the morning after the officials' *Nightline* appearance. There was no customary greeting; instead, the voice on the other end of the phone said, "I'm angry and embarrassed—they had no idea of what they were talking about." Nothing more needed to be said.

Many experts who have studied the federal government's efforts to combat terrorism have reached the same conclusion. In an extensive congressionally mandated report, "Combating Proliferation of Weapons of Mass Destruction," a distinguished commission chaired by former

CIA director John M. Deutch concludes that "the US Government is not effectively organized to combat proliferation."

The Deutch report notes that Congress and the executive branch are often at odds over how to handle proliferation issues. Congress tends to think of the executive branch as indecisive and not interested enough in the threat; the executive branch tends to see Congress as approaching these issues with simplistic solutions and a preference for punitive measures. Congressional action is further complicated by the fact that authority over these issues is shared by at least twenty committees.

Within the executive branch, the situation is even more complicated, the commission report states. "Many offices and agencies have a role in countering the proliferation threat. These include several White House offices; traditional national security elements in the Intelligence Community and at the Departments of State, Defense, and Energy (including the national laboratories); as well as the Departments of Justice, Commerce, Treasury, Health and Human Services, and Agriculture." The commission, which gives short shrift to the role of HHS, chides that "neither the President, Congress, nor any executive branch official knows how much the various agencies have spent on these efforts or how much they plan to spend in the future. . . . With no one specifically in charge of all proliferation-related efforts, no one is ultimately accountable to the President and to Congress. Thus, the present system lets agencies protect their perceived institutional interests rather than fully contributing to an overall plan for achieving broader objectives.

Blame can be deflected and diffused [to] other participants in the interagency process. Such diffuse responsibility invites inefficiency and ineffectiveness, and avoids accountability."

The institutions protect themselves instead of us, the commission reports: "For example, nearly two-thirds of what DoD reports as WMD technology investment for fiscal year 1999 is in missile defense, a traditional war-fighting requirement that would exist even without a WMD threat. A similar tendency toward traditional roles and missions can be found in the Department of Energy's reported WMD technology program."

Similarly, a General Accounting Office report issued in May 1999 took withering aim at "issues to be resolved to improve counter-terrorism operations" and found that the various agencies involved have a long way to go. While recognizing that "federal agencies have conducted several successful interagency operations overseas" in the previous three years, GAO said those agencies nonetheless had not completed the "guidance" documents necessary to make interagency cooperation work routinely. They also had not "resolved command and control issues"—the fundamental questions of who's in charge in a crisis. For example, the FBI had not coordinated domestic guidelines with the Department of the Treasury, which "could have a significant role in an actual terrorist incident." Even more important, international guidelines have been delayed "because the Department of State, the Department of Justice, and the FBI have not reached agreement on the level of State participation in highly sensitive missions to arrest suspected terrorists overseas."

Things have gotten to the point that administration officials testifying before Congress sound almost apologetic—and defensive. "As I sit here today, I cannot tell you that the nation is prepared to deal with the large-scale medical effects of terrorism," Robert F. Knouss, director of the HHS Office of Emergency Preparedness, testified in September 1999. "But we are working very diligently to prepare local medical systems and public health infrastructures, enhance the national health and medical response, and provide for a national pharmaceutical stockpile. There is no 'silver bullet.' "

The programs that are in place show a great deal of money being thrown at the problem—but with priorities that seem to be determined more by which federal agencies are best at budgetary brinkmanship than by where the money might do the most good. The result is big money for medicine—but even bigger money for everyone else.

HHS, in fiscal 1996, got $7 million for its fledgling bioterrorism initiatives, out of $5.7 billion that the rest of the government got in antiterrorism funding. In fiscal 2000 that department received $238 million—out of a total government counterterrorism budget that will probably top $10 billion. Clinton requested nearly $1.4 billion to combat chemical and biological terrorism. "We have to be ready to detect and address a biological attack promptly, before the disease spreads," Clinton said in January 1999.

That $10-billion budget request included only $43.4 million for research and development against biological terrorism, with $40 million to develop and procure drugs and vaccines for the critically important national medical stockpile. The National Institutes of Health will get $24 million for research on diagnostics, vaccines, and other treatments. In all, the budget for HHS was approximately $238 million, a mere $41 million of which could be used to increase public health preparedness in the fifty states and major metropolitan areas.

And still, HHS official Margaret Hamburg says, people began asking her, "What are you going to do with *all that money*?" She says that she's got to make it stretch an awfully long way—and the HHS amounts paled in comparison to the billions in the total counterterrorism budget. The $120 million that was received by the CDC, as part of the HHS budget, hardly begins to address the problem.

Meanwhile, other agencies were going on a real spree.

- The National Guard plans to train and equip 27 full-time "Civilian Support Teams," with 50 more teams to be established in the future. Each team consists of 22 individuals. The cost for 1999 was $52 million. No one is even sure what these teams will do or how they will assist state and local officials with an explosive, chemical or biologic event. A June 1999 GAO report concluded it is unclear how these teams fit into the federal response. Even officials at the FBI and Federal Emergency Management Agency said they do not see a role for these teams.
- The Justice Department in 1999 began handing out $69.5 million in grants to all fifty states and the large municipalities to buy protective gear and detection devices for chemical and biological agents.

- Justice and Defense, along with FEMA, had
  nearly $80 million to hand out to train fire
  fighters, EMS personnel, and other first
  responders.

- The Defense Advanced Research Projects Agency
  (DARPA), the people who brought you the
  Internet, are also getting a healthy chunk of bio
  money. The Clinton administration has asked to
  boost the agency's biological warfare research
  budget 70 percent, to $146 million a year. It is
  awarding grants to develop microbe-killing skin
  lotion, long-lasting antibodies made out of plastic,
  and an alarm system that would circulate in the
  bloodstream and give the "host" a taste signal
  when bioweapons are detected. DARPA officials
  say they know they are making risky choices—so
  much so that the journal *Science* lampooned the
  program with a headline reading, "Too Radical for
  NIH? Try DARPA."

- Texas Tech University in Lubbock is crowing
  about a $7-million appropriation to create a
  national research center for biological terrorism
  issues—part of a $15-million program in
  conjunction with the University of Texas and
  the University of Florida. That money will go to
  creating, among other things, a $2.5-million
  high-performance computer system that will use
  three-dimensional video visualization technology
  "to create simulations of terrorist attacks,"
  according to press reports. (You might be forgiven
  for thinking of it as a multimillion-dollar video

game.) Another $300,000 will go to handheld
devices compared to a "*Star Trek*-like tricorder
that would scan potential victims to see whether
they were exposed during an attack." David
Schmidley, vice-president for research and
graduate studies at Texas Tech, says that the
work is in line with research already going on,
on the effects of pesticides and chemicals found
at Superfund sites: "We will just extend that
capability into the bio-terrorism arena."

Most of these funded projects will have *no impact* on our
ability to respond to a bioterrorism event.

For much of the war on terrorism, HHS has not
even been consulted. The landmark Nunn-Lugar-Domenici
counterterrorism legislation threw funds at the military.
Though the budget planners talked about preparing for
"chem-bio" attacks, in fact their preparation was almost ex-
clusively on the chemical side, and still focused on responding to an obvious attack like the sarin gas in the Tokyo
subway. Funds to make the nation's health system more
ready to respond to biological attacks never seemed to make
it into the plans, says Hamburg of HHS, and work in getting
the health care side of counterterrorism was delayed. "That
was a critical missed opportunity," she says. She recalls
her incredulity listening to government officials speak with
blithe ignorance about defending against chemical attack—
one FBI agent actually said it might be possible to "defuse
the pathogens," as if they were a bomb waiting to go off.

As the probable impact of biological terrorism has

become better understood over time, however, HHS has begun to find itself "at the table" and happy to be there, Hamburg says. However, "It still is a struggle to make sure that our agency is at the table—and to make sure they understand that they still don't get it."

Some of the Nunn-Lugar money did go to building medical "strike teams" that could be brought in for quick response to a crisis, but none of that money went to "people in the white coats," the doctors and technicians who will be facing the patients. Hamburg says that as a top health official in New York City, she'd never heard of the strike teams. "If I was health commissioner in the largest city in the country and had already had a terrorist attack, and if I was unaware of one of our emergency preparations, then we weren't getting through. We weren't developing the right program." Still, she says, other government officials were thrilled with the program; she characterizes their thinking as, "If there's a telephone number to call, you've done your job." Hamburg, of course, disagrees: "Not if no one calls it!"

Despite the new prominence of HHS, Hamburg notes, its links with law enforcement and the intelligence community are still tenuous, and if she needs to take part in a secure videoconference, she has to leave the building. HHS doesn't have the facilities for it; she has to walk down the street to FEMA. For a time, the only secure fax machine for HHS was out in Rockville, Maryland, miles from the downtown-D.C. Hubert H. Humphrey office building. But the budget request for fiscal year 2000 is some $230 million, and Hamburg says, "We've made progress.

We're much more at the table than we used to be." An HHS official even sits in at the National Security Council now, advising on the public health repercussions of national security issues.

Known as Peggy, Margaret Hamburg is another one of the people who prove on a daily basis that the government isn't made up of just pork-barrel fighters and ass-coverers. The nameplate on her desk in Washington reads:

DR. MARGARET HAMBURG, MD
Czarina

It's the kind of joke that suits Peggy, a thin woman with flowing, dark hair and a ready smile. She's a great asset to the federal effort, and I think that much of what has actually been accomplished can be attributed directly to her. But I've seen too many good scientists and administrators in the federal bureaucracy start out as sharp swords and get beaten against rocks for so long that they become dull blades. Officials like Peggy are really trying. But they take a step forward and get pushed back two.

Before getting the job as HHS assistant secretary for Planning and Evaluation, she served as commissioner of health for the city of New York during the World Trade Center attack. She has responsibility for preparing the public health community for bioterrorist attack, a task that sounds downright Sisyphean to me.

We can't even coordinate the roles of the various federal and state health agencies today in the case of simple foodborne outbreaks. Here's an example that still makes me burn: the 1999 *Listeria* outbreak in Bil-Mar cold cuts. The CDC, the Agriculture Department, and regional and local health officials wrangled over authority while dragging their feet on identifying the cause of the outbreak and taking action to contain it. Strong warnings were not issued, and people died because of the government's inaction. I see this happen all the time.

Now imagine the far more chaotic and complex situation following a bioterrorist attack. How could we expect coordination when you've added law enforcement, intelligence, and even the military into the mix? It will take a lot more training and real-life drilling, not tabletop exercises, to get us where we need to go.

One of the best-of-breed federal programs funded in the wave of Washington interest in bioterrorism combines the military's skill at grantsmanship with the need for getting information to the frontline health care providers. Developed by the U.S. Army Medical Research Institute of Infectious Diseases (USAMRIID), the $1-million program, now in its third year, is a three-day video broadcast and teaches the basics in recognizing and treating victims of biological weapons. In 1998 approximately eighteen thousand people took the mini-course; in 1999 the number neared thirty thousand. That brings the cost of the course down to less than thirty dollars per student — "a lot of bang for the buck," says the head of the program, Dr. Edward Eitzen of USAMRIID. And although the early version of the program focused largely on the biological warfare and battlefield scenarios, the 1999 version featured an entire day on civilian bioterrorism defense sponsored by the CDC.

Eitzen is not one of those who believe that nothing is being accomplished against bioterrorism. "My personal opinion is that there's a lot being done," he says. Having worked in the field for more than seven years, he says that since 1998, "the train is starting to pick up speed. . . . There is progress being made—it's not all doom and gloom."

I wish that I were as optimistic as Ed. I don't see the actual planning and drilling that civilians will need in order to respond to an event. And many of those thirty thousand watching the program are military health care workers around the world, who aren't part of the local U.S. health care system that will have to respond to an outbreak. The biggest problem, though, is that three days is not much to cover so much information, especially when there's no classroom monitor to rap knuckles for the distractible students. Observed from a viewing room at the FDA building from which the show was recently being broadcast, two of the dozen viewers on-site were taking notes avidly, but three were dozing. These are the first responders, the first line of defense. It made me think of John Bartlett's horrifying experiment in Baltimore. If he came up dry with his hypothetical anthrax cases at what might be the best medical center in the world, what can we expect from doctors scattered around the world who got a three-day video course?

The same problems of coordination and confusion that can be found across the federal government when it comes to biological terrorism are just as pronounced in the vertical

relationships between the federal, state, and local governments. As I pointed out in the preceding chapter, questions of legal authority are far from resolved, and various
laws offer conflicting guidance. Links between the federal
and law enforcement and intelligence communities and
their local counterparts tend to be tenuous at best, says one
intelligence official, but "they have to share information" to
be effective against terrorist threats. Local police have the
best chance of spotting something odd in their own communities, the intelligence official says, but few local police
departments have invested in special investigations teams
capable of engaging in counterintelligence. Instead, most
police work involves reacting to calls. "Law enforcement
needs to be trained in how [biological terrorism] works,
what the most likely methods of attack are—so that they
will recognize the M.O., they will recognize the delivery
system."

Another big part of training that needs to be established
before an attack is to get local police to understand their
crucial role in preventing chaos—even if that means ne-
glecting family needs in a crunch. "If you're the local cop
and a terrorism event takes place—everyone's coming down
with smallpox—are you going to do your job, or are you
going to say, 'Hey, I'm going to go home and protect my
family'?" If a breakdown of society can be prevented, it
will rest in no small part on their shoulders, the source argues. He says that the police need to be told from the start
that "if you're going to be in this position, this is what has
to be done when the time comes." Yet at the moment, he
contends, the degree of preparation at the local level for
that kind of situation is "zero. You talk to people about

this and they say, 'Yeah, right. That will never happen here.' People go into denial, or say it's somebody else's job."

The problem, this intelligence source suggests, is a leadership catch-22: local authorities are unwilling to spend time and effort on programs without federal support, and Washington has yet to act decisively. Or, as he more pungently put it: "The bottom is waiting for the top to tell 'em what to do—but our political leadership is totally gutless."

9. MITIGATED DISASTER

Up the lake from Chicago lies Milwaukee. A fair number of the shoppers along the city's famed Miracle Mile of stores on any given day comes down from the town just as Kristen Schafer did the day after Thanksgiving. The visitors to Water Tower Place carried the terrorist's weapon back, unwittingly, within their own bodies. And, as in other cities and states, the cases start to appear within two weeks of the attack.

But there's a difference. While the response in Chicago has been haphazard and panicky, officials in Milwaukee and the surrounding region had planned for the worst, taking part in training programs and exercises. That city got a harsh reminder of what an infectious agent could do after more than 400,000 of its citizens developed a miserable diarrheal disease from contaminated drinking water nine years before. So as soon as word gets out that smallpox has been loosed on the world, the mayor calls a meeting with the Milwaukee Regional Bioterrorism Working Group. It includes all of the right players—public health, medical care delivery, law enforcement, emergency management, and elected officials. The governor, realizing both the real and the symbolic need to show support, overrides his staff's concern about getting infected and flies into Milwaukee's emergency center by helicopter. The group follows the script of its emergency preparedness plan: initial cases will be placed within specially created rooms in the city's newest hospital that provide negative air pressure and super-fine HEPA filters in the air-conditioning to keep the virus from escaping.

The quarantine effort is carried out with a sense of purposefulness; and with plenty of public information, compliance is far higher in Milwaukee than in Chicago. Public health, not the FBI, is in charge. The panic is there—nothing can stop it—but the city's leadership is winning, keeping the sense of terror from incapacitating everyone. Volunteer police in HEPA filter masks supervise the quarantined town. City officials agree to take over the old city armory and turn it into a vast smallpox center; later they will commandeer a number of heated warehouses as field hospitals as the number of patients grows. Suspected smallpox cases are directed away from hospitals to the makeshift centers in the warehouse district, keeping them from otherwise healthy people who could become infected.

So when Kristen becomes ill, she is quickly isolated and her family is put under quarantine to watch for signs of new illness in the children and her husband. The disease tortures her, but the pain of not seeing her family is almost as bad. Though she cannot see her children, she can pick up a phone in lucid moments and call them. "I'm going to beat this," she promises.

Knowing that chaos is a likely outcome of an epidemic, the governor starts working the phones immediately to call in the National Guard to enforce order in the town.

Most important, the governor uses every bit of influence in his power to get smallpox vaccine sent to the city right away, calling in favors throughout Washington, including both Wisconsin senators. He knows that the task of vaccinating people will be vast—but also knows that he can't do anything unless he gets the limited supplies of medicine for his city.

That night the governor and the mayor place a call to the President. A political protégé and friend of the governor, the President assures them that "they're loading the vaccine onto the plane now." The governor looks up at the ceiling and mutters, "It's hard to believe there's not going to be enough for every city that's been hit by this—but that means we have to make sure we get ours."

The mayor shakes his head. "This is hell, isn't it?"

The governor smirks. "It's life, back to the basics. Us or them. Survival or death. We've got an obligation to our citizens." He lets out a deep sigh. "You know, it's like that old story of the two campers who wake up in the middle of the night because they hear a bear—a hungry bear. One of them quietly puts on his sneakers and begins lacing them up. 'You're crazy,' the other says. 'You can't outrun a bear.'

" 'I don't have to outrun the bear,' he says. 'I only have to outrun you.' "

The mayor and governor host daily live briefings describing

the actions they have taken, explaining the reasons for the harsh police action and explaining to the citizens how to help themselves. "Make no mistake, this is a disaster," the governor tells his citizens in a grim fireside chat. "But it's a disaster that becomes much, much worse if we let things get out of hand. If we all cooperate—and that means taking tough action and sticking to the plan—many, many more of us will survive."

Mass vaccination centers begin working right away, with vaccinated volunteers wielding the oddly shaped bifurcated needles used to scratch the smallpox vaccine into the skin, fifty thousand vaccinations each day. By the end of the first week, food deliveries have gotten under way for homes under quarantine and police vehicles are providing transportation to the smallpox center.

The differences between Milwaukee and Chicago don't really show at first: people are dying by the hundreds, and then by the thousands, in both cities. But after four weeks, the difference in planning and execution is apparent. In Milwaukee the quick quarantine work and the steady administration of vaccine to medical crews and then to tens of thousands of citizens limit the size of the second wave of patients and prevent a fourth wave from emerging at all. Make no mistake, the governor's ability to politically wrestle a substantial amount of vaccine to Milwaukee is key. The number of cases levels off at a horrifying 5,000, with some 1,300 deaths—a catastrophe by any measure except one. Down the blockaded highway in Chicago, where the epidemic still rages, the number of deaths has already climbed past 140,000.

• • •

Months later the repercussions are still being felt. Milwaukee is rebuilding. As hard as the crisis was, the people of the city say that the problems have brought them together. The governor appoints a high-profile panel to build on the preparedness efforts and to help other states to do the same. He also calls for a volunteer corps of vaccinated health care workers and police, as well as smallpox survivors, as a quick-response team for other parts of the country that might get hit— "the bulletproof brigade," he calls them.

As he announces the new organization, he introduces its new director: the spotlight shines on a woman whose face is pitted with smallpox scars; a single child stands beside her on the dais. "This woman went through the flames," the governor says, "and she came out on the other side. She knows the pain this disease can cause and knows that hope can help to beat it. Ladies and gentlemen, it is my great honor to be able to introduce to you Kristen Schafer!"

Back in Chicago the dying has finally stopped. But the city itself seems dead, its citizens past hope or caring. Large businesses that transferred operations to other cities temporarily during the crisis find little reason to return, and despite proclamations by the mayor that "we will rebuild," the city sinks further into economic depression and lethargy. Those with the ability to do so move away, but it's hard to move when no one is buying houses. Thousands default on their mortgages, leaving the house keys in the mailbox and driving away.

Yuri was happy to leave his seedy apartment building and the city he had come to despise. He has found a place much

more to his liking, a Sunbelt haven. The payment for his ser-
vices goes a long way here—his apartment is larger, airy, and
bright. There is little furniture: a single chair in the living room,
in front of a big-screen television. He rarely turns it on. He has
no patience for sitcoms or the bathos that passes for drama
on television. But he knows he'll be firing it up before too
long: he will want to watch the news bulletins of his next job
in style.

READERS LOOKING FOR a happy ending might be dis-
appointed by what they find here. There is none to offer,
no stunning dénouement, not even a punch line. Instead, I
present recommendations on the best way to climb out of
the hole that our nation has dug for itself. If you think
there's a simple solution, you're wrong. Nor is there a
comprehensive plan that will perfectly prepare us for a
biological terror event—for no matter what we do, we
can't prevent the approach of the invisible man. But given
the catastrophic potential of bioterrorism, we must face
the urgent need to address many yet-unattended issues.
And every private citizen has an obligation to hold elected
officials responsible for addressing those issues.

Whatever we do, America will remain a uniquely compelling target for terrorists. But our lack of preparedness
doubtless heightens our vulnerability to bioterror attack.
So far, most of what we have done has been to react to incidents; now it is time to act, to prepare. Although the law of
diminishing returns understandably limits what actions we
should take in the name of prevention, there are neverthe-
less a few things that must be done—at the federal, state,

and local levels, and by each of us—to make a difference. We can take steps now that could both make it harder for terrorists to commit evil and keep the damage they do from growing out of control. And we can steer the outcome of an attack from the unmitigated disaster of the Chicago scenario to the "mitigated disaster" of Milwaukee.

Below, my eight-point plan for change:

1. STOP TALKING ABOUT "WEAPONS OF MASS DESTRUCTION." I'm not talking about conveniently erasing these weapons out of our everyday world, though it would be a miracle if such magic actually existed. No, I simply mean it's time to stop lumping all weapons that can kill large numbers of people under the single rubric of WMD. The difference in responding to bioterrorism, as opposed to a chemical or nuclear attack, is like the difference between flying a plane and driving a Formula One car. Both are moving vehicles, but very different skills are required for each one. The overuse of the term "weapons of mass destruction" has done a great deal to stunt the necessary attention to the looming threat of biological terrorism. It has allowed policy makers to throw money at the broader problem, shortchange this narrower one, and still claim to be solving the problem. As we've seen, in contrast to other forms of WMD, bioterrorism response is not primarily a military and law enforcement effort. It's a public health and medical system effort.

I don't actually expect the phrase to go away, any more than I expect the weapons to. Buzzwords are like viruses, neither alive nor dead but moving from host to host, seemingly forever. But we should all insist that policy makers acknowledge that biological terrorism is different from terrorism that relies on chemical weapons or explosives, and deserves separate consideration. That means our budgets at the federal, state, and local levels have to show proper funding for bioterrorism planning, training, monitoring, and stockpiling. In 1999, the CDC supported funding of $41 million for all fifty states and three large metropolitan areas—a minuscule amount in light of the $10 billion spent on terrorism. Yet those public health and medical programs are our first, second, and third lines of defense against and in response to a biological weapons attack. To put it bluntly, our priorities are really screwed up.

Our laws should be rewritten to recognize the distinction between responding to most weapons of mass destruction and responding to a bioterrorism attack. Terry O'Brien's analysis of the gaping holes in our legal system shows issues that must be addressed before a crisis, not during one. Otherwise, when we finally do have to authorize and carry out a quarantine, valuable time will be lost figuring out who is in charge and sorting out issues of legal authority. In a bioterrorist event, loss of such time will translate directly to loss of human lives; to prevent this, I believe that the administration and Congress should appoint a bipartisan national legal panel to draw up model legislation and enact it as quickly as possible.

Removing the WMD bias is most important in the area that policy makers call consequence management—running the show in the aftermath of an attack. I hope I have made the point that responding to a biological attack requires an entirely different structure and management system than responding to a chemical or bomb attack. At the moment, coordination of response to WMD attacks falls to the Department of Justice and the Department of Defense. To be sure, that is the right management team for a blast or chemical release: the cops and soldiers should remain the go-to guys in that kind of crisis. But you don't want them running the show during a biological attack, any more than you would expect them to coordinate the response to an outbreak of listeriosis at a hot dog plant, Legionnaires' disease from a cooling tower, or even West Nile virus in New York City. Those crises require special skills, special knowledge, and special people—all already present within the public health system. The Centers for Disease Control and Prevention has been late to recognize its potential role in biological terrorism response, and its leadership may have room for improvement, but since 1999 it has become a more active participant in the process and should be placed in charge of civilian biodefense.

2. BUILD THE STOCKPILE. Until we have a large and usable stockpile of the right antibiotics and vaccines for the most likely agents to be used in a biological attack, we're dead. Nothing can move forward until we have created this fundamental buffer between us and the abyss. Experts have been pushing for a new smallpox vaccine for three years, and seem little closer to having one than when they started. Both the administration and Congress must accept blame for a situation that has shown the worst of the federal bureaucracy. Yes, creating a stockpile involves a guessing game: a determined terrorist could well find out what agents the stockpile defends us against and hit us with an alternative. But if it means that we're able to respond quickly to an attack of anthrax or smallpox, it is well worth the effort. And yes, it will be expensive, but just a fraction of what we are currently wasting on other terrorism preparedness schemes today. It's part of the essential reordering of priorities that goes with rethinking "WMD": we must have fewer tricorder contracts and a lot more vaccines.

3. BUILD MORE "SURGE CAPACITY."   At the moment, hospitals, pharmaceutical companies, and insurers squeeze every excess penny out of health care, performing at the limits of their capacity. It's time to open the debate over how much we're going to let economics be the single compass for directing our medical system; we need, as a nation, to build a little more slack into the system. The added capacity would have the side benefit of better preparing our health care networks for natural disasters and the still-possible pandemic of influenza like the one that carried off so many millions of people worldwide at the beginning of this century. It also will be expensive—but then, so are fire departments at airports. When was the last time the fire department at your nearest metropolitan airport responded to a plane crash? Still, we would never operate those airports without fire fighters on duty twenty-four hours a day, every day of the year. History shows us that we pay for what we think we need, and when we understand how much we need this, I'm confident we will pay for it. If we don't, we'll *really* pay for it.

We desperately need doctors, particularly infectious-disease experts, and nurses to participate in local and regional planning activities for bioterrorism. But they almost never show up. Why? In large part, they are so stretched in their capacities to provide more patient care with less resources, they have no "financial freedom" to spend time at an all-day meeting without some reimbursement to their hospital or managed care organization. Our failure to address this is penny-wise and pound-foolish.

Part of our surge capacity process will involve assembling medical teams to supplement the staffs of local hospitals and treatment centers wherever outbreaks may occur. Prior to any attacks, these professionals, who would come from the ranks of trained medical personnel nationwide, would voluntarily receive the vaccinations they need to be able to go safely into the nation's new hot zones. Although I knew it would be their job to do so, my worst nightmare when I was at the Department of Health was the prospect of forcing my staff to investigate an outbreak of smallpox in my state. As most of them have little if any immunity against smallpox, I knew it would be a death sentence. Military leaders have become experts in the unhappy business of putting people in harm's way and suffering "acceptable losses," but public health officials have not—and won't ever have to if we prepare these medical teams.

We'll also need to enlist the help of mental health professionals equipped to help counsel the survivors, which could include entire cities of people whose world and lives will have been shattered by the advent of the unthinkable.

4. SHORE UP THE PUBLIC HEALTH INFRASTRUCTURE TO BE READY FOR QUICK RESPONSE TO OUTBREAKS.   This point is related to the third item, but goes further and deeper. Along with helping the people who will treat patients on the front lines, we have to strengthen the broader public health system that supports their efforts. The first major phase of the nation's new infectious disease detection program, a nine-site network of monitoring and diagnostic centers (now receiving only $12 million of annual funding), must grow. The $41 million for the CDC's first grants to 53 state and local public health programs must also grow quickly. Current levels provide only very limited resources for any one state or large city, given the potential need. With our public health infrastructure in its current shape, trying to detect and respond to a bioterrorism attack is comparable to running O'Hare Airport's air traffic control system with tin cans and string.

Like the proposed buildup in surge capacity, strengthening our public health system is "dual use" in the best sense of the phrase. The improvements will be felt by the entire nation as we find ourselves better able to detect and combat natural outbreaks like foodborne pathogens, influenza, and the next West Nile virus scare. They will also help us fix problems of our own making such as the rise in antibiotic-resistant strains of bacteria; there's no risk that we'll be spending money just to have a lot of people sitting around idle.

It is this reinforced public health infrastructure that will be able to respond to outbreaks — natural and man-made — by mobilizing the necessary antibiotics and vaccines and getting them to the people that need them. Building an adequate stockpile of vaccines and antibiotics won't mean much if the cache is locked in a vault in Atlanta and nobody can get it to the citizens who need it.

Having to scramble to get antibiotics and vaccines to a large population isn't as rare as you might think it is. Remember the meningitis outbreak I discussed earlier, where our team was stretched to the breaking point with a need to distribute vaccine and antibiotics to only thirty thousand people. It occurred under the watch of one of the best health departments in the country—and it stretched us to the very limits of our ability. Now imagine needing to vaccinate millions of people!

5. Clear Up the Roles of Federal, State, and Local Governments.   Just as we need to define the roles of the various agencies across the federal government, we need to drill down through the layers of bureaucracy and clarify the roles and responsibilities on the state and local levels. Our efforts to turn around the lagging preparedness issues at the top don't automatically ensure that the same problems will be resolved at the other levels. Local police and medical teams don't have any better understanding of each other than the federal Departments of Justice and HHS do, but the federal government can help by setting a better example. Heads of federal agencies, too, can improve matters by treating the funding for biological terrorism as less of an opportunity for pork-barrel grantsmanship and more of an opportunity to help the nation head off catastrophe.

This requires leadership at every level of government. Congress has held hearing after hearing on issues of WMD terrorism over the past four years, but what leadership have they contributed? I would say very little; so far, they have added to the confusion by breaking up the field into the jurisdiction of countless committees and by providing categorical funding that doesn't guide the federal agencies to do the best work they possibly can. State officials also need to show adaptability and leadership. They invite disaster by taking the easy route and leaving these issues entirely to federal lawmakers. Like their Washington brethren, they aren't acting, only reacting. But state lawmakers can also plan ahead, funding local training programs and beefing up the disease surveillance capabilities of their own state health departments. Many states prepare their local health and law enforcement professionals for natural disasters like earthquakes and fires, discrete events that do their considerable damage in a definite span of time; it shouldn't be hard to see the value of preparing for a manmade disaster that could cause the full-scale economic collapse that a large outbreak of contagious disease could cause.

6. CLEAN UP THE COVERAGE.   Most of the press coverage of biological terrorism has been made up of scare stories, the give-'em-the-gross-details writing we like to call gorenography, and gee-whiz pieces detailing the high-tech schemes that various agencies are funding. That's a shame, because thoughtful news coverage could help keep lawmakers and agencies focused on the problems at hand, and keep them honest besides. That's the role of the press envisioned by the authors of the Constitution—as key players in the national marketplace of ideas. First Amendment protection was granted to the press because the questions that journalists ask were seen as an essential part of the machinery of democracy itself. Instead, we're inundated with celebrity gossip and daily handicapping of political horse races. Today's press serves the attention-deficit generation, not the needs of the nation.

A few reporters have focused on the issues of biological terrorism intelligently, and with a critical eye: Laurie Garrett's work for *Newsday* comes to mind, as does Richard Preston's work for the *New Yorker*. David Kaplan at *U.S. News & World Report* and Judith Miller and William Broad at *The New York Times* also shine a light in areas that desperately need to be seen. A single story doesn't shift the direction of the ship of state—though Preston's chilling July 1999 report on smallpox should have!—but the information that top journalists like these put before the public helps inform us all and should lead to better policies and programs.

Reporters and editors also need to prepare themselves for writing about these outbreaks by learning what they can about the diseases that might be used. Reporting inaccurately that anthrax is a communicable disease like smallpox could worsen the panic in the midst of an attack. Journalists aren't agents of the government, and shouldn't be. But journalism, at its best, does serve the public interest.

7. WE'LL UNDERSTAND IT IF WE ACTUALLY PRACTICE.   Most everyone can recall seeing a picture in the newspaper or video footage of the classic WMD exercise. Typically, a number of HAZMAT professionals are seen in space suits walking out of some building carrying a container. We all feel comforted to know that the government has an impressive effort for terrorism. The painful irony is that these exercises do nothing to prepare us for the eventual bioterrorist attack.

As I noted before, we have fooled ourselves into believing we're prepared to deal with bioterrorism because we have perfected our response to an event such as an explosion or release of a chemical agent. In real life, none of these players, including the FBI or other law enforcement officials, will be on the front lines when we recognize the results of the intentional release of a biologic agent. Moreover, that recognition will occur not over minutes to hours, but rather over days to weeks. In the end, it will be the emergency rooms, doctor's offices, and public health departments that will be the smoke alarms going off alerting us to the impending raging fire.

Despite this conclusion, we continue in this country to avoid preparing for bioterrorism through such activities as meaningful live drills and tabletop exercises (a type of make-believe exercise usually conducted in a single room). Why? Frankly, to unfold a bioterrorism exercise that is realistic means days to weeks of challenging health care workers, persons working in clinical laboratories, and public health officials with bits of information that appear to be unrelated. And it won't happen in a single clinic, hospital, or even geographic region. Most of all, no one will even know it happened. That's different from responding to a recognized crisis, even if you don't know why the building blew up.

For these reasons, very few communities have attempted to play out realistic scenarios involving the release of a biologic agent. Instead, we continue to fall back on exercising the classic chemical release to earn the comfort of knowing that our HAZMAT teams are in place. This is a serious mistake. We need to begin to organize, on a regional basis, plans for addressing head-on the complexities of a one-week to several-months scenario that could mean simulating the provision of antibiotics and vaccine to hundreds of thousands of individuals and direct medical care for an equal number of critically ill patients.

These types of drills will take resources. Unfortunately, both the public health and medical care delivery systems are already stretched to the point of breaking by their efforts to provide the necessary resources for day-to-day business. There isn't any flexibility in these systems to allow for the kind of exercises that will allow us to understand and address the serious deficiencies in our bioterrorism response protocols. In addition, state and federal planning efforts to date have generally neglected hospitals. While first responders, EMS, and law enforcement have become very energized about this issue, there has been very little attention paid to what needs to be done within the hospitals.

The threat of bioterrorism raises many difficult questions for hospitals. So far, the most ambitious attempt to address these issues is *Bioterrorism Readiness Plan: A Template for Healthcare Facilities*, a report prepared by the Association for Professionals in Infection Control and Epidemiology (APIC) and the Bioterrorism Working Group in the

Centers for Disease Control and Prevention. The report, which runs thirty-four manuscript pages, provides general recommendations for responding to a suspected bioterrorism event for a hospital. While noting that hospitals need to prepare their plans in collaboration with local and state health departments, the report has been criticized for not elaborating on the need for regional planning to coordinate actions by multiple health care facilities and other agencies in response to a major biological attack. I believe that no health care facility should consider itself an island in planning for things such as outbreak detection, patient placement and transport, discharge management, and postmortem care. Yet today many health care systems within a given metropolitan area are highly competitive and have very little in the way of joint communication for efforts such as bioterrorism planning. While a single hospital can begin to address the issues of how it will respond to a massive influx of potentially infectious patients and the protection of its workers, much of this will be of little use if the availability of antibiotics and vaccines are dictated by others outside the hospital system.

If we really want to begin to address local, regional, and even national strengths and weaknesses in our bioterrorism response plan, we've got to become more realistic about how we prepare for such events. Drills based on realistic scenarios that involve all of the players that will ultimately be involved in a real bioterrorism event must take place. While HAZMAT teams will play critical roles should a chemical terrorism event occur, let us not be lulled into a false sense of security by seeing their impressive gear when we address the issue of bioterrorism.

8. WE'RE ON OUR OWN—TOGETHER. What does this leave for individuals to do? Plenty, actually. Citizens need to keep informed about what is being done in their name and to think about whether the things that are being done truly serve their interests. Then take that knowledge and use it to pressure our elected representatives at the federal, state, and local levels to do the right thing, fund the right programs, and make sound choices for the future. Each of us has to demand more accountability of our elected officials—and not to confuse performances on *Nightline* with performance of their duties.

You might expect me to advise you to get vaccinated against the most likely diseases to be used in biological terrorism. I won't, though—because it's the wrong thing to do. Yes, we'll need the vaccines and antibiotics for the outbreaks, but not as a part of a routine program.

It goes against the simple realities of statistics. No individual in America is highly likely to be infected by a biological terrorism attack, which after all will affect only those directly exposed—or, in the case of contagious diseases, those who come into contact with the initial victims. This means that the likelihood of being exposed to one of these agents for any single American is quite low, kind of like getting struck by lightning. Moreover, getting protected against anthrax requires up to six shots, and the current smallpox vaccine has side effects that would be unacceptable to many people today, especially in light of advances made in producing vaccines with far fewer side effects for other diseases.

I worry that disease hustlers will begin encouraging people to pay top dollar to be vaccinated against anthrax and smallpox as moneymaking schemes, pitching their wares to the worried well. Marketers say that sex sells, but sex doesn't have anything on fear. Don't give in to the hype. The appropriate use of these vaccines will be in association with an outbreak, or in advance for a limited number of volunteer public health and health care workers, police, and other personnel needed to maintain our basic infrastructure support during the crisis.

Ultimately, the lesson of this book is that we can't take bugs for granted anymore. Terrorists are acting as intermediaries to bring the problem to us, but we've been reminded again and again, with outbreaks like antibiotic-resistant TB, HIV, *E. coli* O157:H7, and West Nile virus. I'm not just talking about biological terrorism. We are all fighting a much bigger war: the eternal evolutionary battle between man and germ. The bugs were here before we were here and the bugs will be here after we're gone. But we have to learn the ways of the adversary, fight on his terms—and survive.

I pray that I'm wrong about the likelihood of a bioterrorist attack, that it never happens. But if I am not, we will all be judged in the terrible aftermath by how well we were prepared.

Reading through these chapters about the evil that men do, and the worse evil that they may do in the future, you might come away with a feeling of hopelessness about the human condition. But despite the forcefulness of our warnings, I remain positive. I choose to view these modern plagues as a kind of test of the human spirit; for no matter what happens, when confronting the worst, we will see many people give their best.

Existentialist novelist Albert Camus understood this well. At the end of *The Plague*, his brilliant novel set within a modern disease outbreak, he concludes with an elegant testimonial to the essential goodness within us, and the recurring struggle against dark forces in our hearts and within biology itself. The novel closes with the main character, Dr. Rieux, contemplating the celebration marking the end of the outbreak:

> And it was in the midst of shouts rolling against the terrace wall in massive waves that waxed in volume and duration, while cataracts of colored fire fell thicker through the darkness, that Dr. Rieux resolved to compile this chronicle, so that he should not be one of those who hold their peace but should bear witness in favor of those plague-stricken people; so that some memorial of the injustice and outrage done them might endure; and to state quite simply what we learn in time of pestilence: that there are more things to admire in men than to despise.
>
> Nonetheless, he knew that the tale he had to tell could not be one of a final victory. It could be only the record of what had to be done, and what assuredly would have to be done again in the never ending fight against terror and its relentless onslaughts, despite their personal afflictions, by all who, while unable to be saints but refusing to bow down to pestilences, strive their utmost to be healers.

And, indeed, as he listened to the cries of joy rising from the town, Rieux remembered that such joy is always imperiled. He knew what those jubilant crowds did not know but could have learned from books; that the plague bacillus never dies or disappears for good; that it can lie dormant for years and years in furniture and linen-chests; that it bides its time in bedrooms, cellars, trunks, and bookshelves; and that perhaps the day would come when, for the bane and the enlightening of men, it would rouse up its rats again and send them forth to die in a happy city.

NOTES

INTRODUCTION

xvii . . . *keeps me awake at night:* Judith Miller and William J. Broad, "Clinton Describes Terrorism Threat for 21st Century," *New York Times,* January 21, 1999, p. A1.

xviii . . . *more threatening than either explosives or chemicals:* D. A. Henderson, "Bioterrorism as a Public Health Threat." *Emerging Infectious Diseases,* vol. 4, no. 3, (July–September 1998), National Center for Infectious Diseases, Centers for Disease Control and Prevention.

CHAPTER 1

2 . . . *tough little pod:* Peter C. B. Turnbull, "Bacillus,"
Chapter 15 of *Medmicro:* online reference work.
http://129.109.136.65/microbook.ch015.htm.

3 . . . *MAPKK:* Phillip Hanna, Nicholas Duesbery,
George Vande Woude, Stephen Leppla, "How
Anthrax Kills," *Science,* June 12, 1998, p. 280.

5 . . . *Plague, caused by the bacterium* Yersinia pestis, *can
be traced to the earliest records:* Thomas W. McGovern,
Arthur M. Friedlander, "Plague," *Textbook of Military
Medicine,* Part One, "Medical Aspects of Chemical and
Biological Warfare," Chapter 23 (Office of the
Surgeon General, 1997).

15 . . . *Smallpox, the nightmare to end all nightmares:* D. A.
Henderson, "Smallpox: Clinical and Epidemiologic
Features," *Emerging Infectious Diseases,* vol. 5, no. 4
(July–August 1999); also, D. A. Henderson, Thomas V.
Inglesby et al., "Smallpox as a Biological Weapon:
Medical and Public Health Management," *Journal of the
American Medical Association,* vol. 281, no. 22 (June 9,
1999) pp. 2127–2137; also, F. Fenner, D. A. Henderson,
I. Arita, Z. Jezek, I. D. Ladni, *Smallpox and Its
Eradication;* World Health Organization; also, *Textbook of
Military Medicine,* Part One, "Medical Aspects of Chemical
and Biological Warfare," Chapter 27 (Office of the
Surgeon General, 1997).

19 . . . *Anthrax is a brilliantly efficient killer:* Thomas V.
Inglesby, D. A. Henderson et al., "Anthrax as a Biological
Weapon, Medical and Public Health Management,"
Journal of the American Medical Association, vol. 281. no. 18
(May 12, 1999), pp. 1735–1745.

20 . . . *plague is a disease that has evoked panic and fear in populations dating back to our earliest history:* Thomas W. McGovern, Arthur M. Friedlander, "Plague," *Textbook of Military Medicine,* Part One, "Medical Aspects of Chemical and Biological Warfare," Chapter 23 (Office of the Surgeon General, 1997).

21 . . . *Botulism is typically a foodborne ailment:* Stephen S. Arnon, Robert Schechter et al., "Botulinum Toxin as a Biological Weapon: Medical and Public Health Management," *Journal of the American Medical Association,* in press; also, "Biological Warfare and Terrorism: The Military and Public Health Response," student material for satellite broadcasts (September 21–23, 1999), U.S. Army Medical Research Institute of Infectious Diseases.

21 . . . *There are a few forms of the bacterial disease tularemia:* "Biological Warfare and Terrorism: The Military and Public Health Response," student material for satellite broadcasts (September 21–23, 1999), U.S. Army Medical Research Institute of Infectious Diseases; also, Ken Alibek and Stephen Handelman, *Biohazard: The Chilling True Story of the Largest Covert Biological Weapons Program in the World — Told from Inside by the Man Who Ran It* (Random House, 1999).

22 . . . *Hemorrhagic fevers are the horrifying diseases:* Peter B. Jahrling, C. J. Peters, et al. "Viral Hemorrhagic Fevers as a Biological Weapon: Medical and Public Health Management," *Journal of the American Medical Association* (in press); also, "Biological Warfare and Terrorism: The Military and Public Health Response," student material for satellite broadcasts (September 21–23, 1999), U.S. Army Medical Research Institute of Infectious Diseases;

also, Ken Alibek and Stephen Handelman, *Biohazard: The Chilling True Story of the Largest Covert Biological Weapons Program in the World — Told from Inside by the Man Who Ran It* (Random House, 1999); also, "Global Proliferation of Weapons of Mass Destruction," hearings before the Permanent Subcommittee on Investigations of the Committee on Governmental Afairs, U.S. Senate, October 31 and November 1, 1995.

CHAPTER 2

29 . . . *just like a pair of dominoes:* "Global Proliferation of Weapons of Mass Destruction," hearings before the Permanent Subcommittee on Investigations of the Committee on Governmental Affairs, U.S. Senate, 104th Cong., 2nd sess., pt. 3, March 27, 1996; Blaine Harden, "2 Guilty in Trade Center Blast; Engineer, Driver Face Life Imprisonment for Fatal Attack," *Washington Post,* November 13, 1997, p. A1.

30 . . . *self-proclaimed apocalyptic prophets at home:* William S. Cohen, "Preparing for a Grave New World," *Washington Post,* July 26, 1999, p. A19.

30 . . . *Death, the unseen Death, is coming:* H. G. Wells, *The Invisible Man* (Berkley, 1968), p. 127.

31 . . . *in furtherance of political or social objectives:* "Terrorism in the United States 1997," Federal Bureau of Investigation, Counterterrorism Threat Assessment and Warning Unit, National Security Division.

32 . . . *You simply want to wipe them out:* Mike Reynolds, personal interview.

33 . . . *crisis of 1962:* Ashton Carter, John Deutch, Philip Zelikow: "Catastrophic Terror: Tackling

the New Danger," *Foreign Affairs*, vol. 77, no. 6 (November/December 1998), p. 80.

33 ... *at least a dozen terrorist groups have expressed an interest:* Statement by Special Assistant to the DCI for Nonproliferation John A. Lauder on the Worldwide Biological Warfare Threat, to the House Permanent Select Committee on Intelligence, as Prepared for Delivery on March 3, 1999.

33 ... *major attacks are also becoming more likely:* Jessica Stern, "The Prospect of Domestic Bioterrorism," *Emerging Infectious Diseases*, vol. 5, no. 4 (July–August 1999).

36 ... *The United States has identified twenty-five countries:* Worldwide Threat Assessment Brief to the Senate Select Committee on Intelligence by the Director of Central Intelligence, John M. Deutch, February 22, 1996.

37 ... *more than a dozen states have offensive and/or biological weapons programs:* "Combating Proliferation of Weapons of Mass Destruction," report of the Commission to Assess the Organization of the Federal Government to Combat the Proliferation of Weapons of Mass Destruction; William S. Cohen, "Preparing for a Grave New World," *Washington Post*, July 26, 1999, p. A19.

37 ... *the agency singled out Iran, Iraq, North Korea, and Sudan:* CIA biannual report to Congress, "Acquisition of Technology Relating to Weapons of Mass Destruction," covering 1 January through 30 June 1998, published February 9, 1999.

38 ... *I had to fire the guy:* Stephen Morse, personal interview.

39 . . . *it's only a matter of time until that changes:*
C. J. Peters, personal interview.

39 . . . *profiled in the* New Yorker: Richard Preston, "The
Bioweaponeers: In the Last Few Years, Russian
Scientists Have Invented the World's Deadliest Plagues.
Have We Learned About This Too Late to Stop It?" The
New Yorker, March 9, 1998.

41 . . . *It's just hard for me to conceive of 4,500 metric tons of
anthrax:* William Patrick, personal interview.

41 . . . *The effectiveness of those Soviet pathogens was proved,
tragically, at Sverdlovsk:* D. A. Henderson, "Bioterrorism as
a Public Health Threat. *Emerging Infectious Diseases,*
vol. 4, no. 3 (July–September 1998), National Center
for Infectious Diseases, Centers for Disease Control and
Prevention; citing M. Meselson, V. Guillemin, M. Hugh-
Jones, A. Langmuir, I. Popova, A. Shelokov et al., "The
Sverdlovsk Anthrax Outbreak of 1979," *Science,* vol. 266
(1994), pp. 1202–1208.

42 . . . *Alibek testified before Congress:* Kenneth Alibek,
prepared testimony before the House Armed Services
Committee Military Procurement Subcommittee and
Military Research and Development Subcommittee,
October 20, 1999.

43 . . . *Iraq had twenty-five missile warheads filled with anthrax:*
"Global Proliferation of Weapons of Mass Destruction,"
hearings before the Permanent Subcommittee on
Investigations of the Committee on Governmental
Affairs, U.S. Senate, 104th Cong., 2nd sess., pt. 2, March
13, 20, and 22, 1996, p. 93.

43 . . . *missile warheads, aerial bombs, and aircraft-mounted
aerosol spray tanks:* "CIA, Iraqi Weapons of Mass

Destruction," February 13, 1998, cited in Glenn E. Schweitzer and Carole C. Dorsch, *Super-Terrorism: Assassins, Mobsters, and Weapons of Mass Destruction* (Plenum, 1998), p. 119.

44 . . . *China as one of "the most serious threats":* "Combating Proliferation of Weapons of Mass Destruction," report of the Commission to Assess the Organization of the Federal Government to Combat the Proliferation of Weapons of Mass Destruction.

46 . . . *he deeply understood the likelihood of such an event:* The late King Hussein of Jordan, private communication.

47 . . . *Monte Kim Miller:* Charlie Brennan, "Israel Detains 8 Adults of Apocalyptic Group," *Rocky Mountain News,* January 4, 1999, p. 5A; Tom Kenworthy, "Relatives Left in Dark About Kin Who Joined Christian Cult Leader," *Washington Post,* January 8, 1999, final edition, p. A3.

48 . . . *they would have to fight with every available weapon:* Jessica Stern, "The Prospect of Domestic Bioterrorism, *Emerging Infectious Diseases,* vol. 5, no. 4 (July–August 1999).

50 . . . *Anybody can bring the battlefield to our porch now:* Michael Osterholm and Mark Olshaker, unpublished manuscript.

51 . . . *might have killed thousands or even tens of thousands:* Stefan Leader, "Osama bin Laden and the Terrorist Search for WMD," *Jane's Intelligence Review,* June 1999.

51 . . . *Fortunately, the group ran out of time:* Information for these passages is drawn from many sources, including "Global Proliferation of Weapons of Mass Destruction;" hearings before the Permanent Subcommittee on Investigations of the Committee on Governmental

Affairs, U.S. Senate, 104th Congress, October 31 and November 1, 1995; also, David Kaplan and Andrew Marshall, *The Cult at the End of the World.* (Crown, 1996); Jessica Stern, *The Ultimate Terrorists* (Harvard, 1999).

53 . . . *We hope to be rewarded for it by God:* Interview, ABC News; on-line at *http://abcnews.go.com/sections/world/ DailyNews/transcript_binladen1_ 981228.html.*

54 . . . *a disjointed, shadowy confederation of extremists:* Colum Lynch and Vernon Loeb, "Bin Laden's Network: Conspiracy of Loose Alliance?" *Washington Post,* August 1, 1999, p. A1.

54 . . . *Islamic Jihad . . . has devised both chemical and biological weapons:* "Jihad Says It Has Chemical and Biological Weapons: Report," *Agence France-Presse,* April 19, 1999.

57 . . . *racial warriors who believe in acting alone:* Hanna Rosin, "Suspect in Community Center Shooting Surrenders, Admits Hatred; a 'Lone Wolf' Priesthood of Aryan Resistance," *Washington Post,* August 12, 1999, p. 1.

58 . . . *That's how we create soldiers:* Mike Reynolds, personal interview.

58 . . . *conditioning helped to raise the "kill rates":* David Grossman, *On Killing: The Psychological Cost of Learning to Kill in War and Society* (Back Bay, 1995), p. 24.

59 . . . *Phineas Priesthood:* Numerous sources, including Jim Nesbit, "Mixing the Bible With Bullets," *Denver Post,* August 22, 1999, p. H6, and Karen Brandon and Michael J. Berens, "Hate Groups Use Recent Shootings as Recruiting Aid," *Chicago Tribune,* August 15, 1999, p. 4.

60 . . . *How many cities are you willing to lose before you back off?* David E. Kaplan, "Terrorism's Next Wave: Nerve

Gas and Germs Are the New Weapons of Choice," *U.S. News & World Report,* November 17, 1997.

60 . . . *Harris reappeared:* Tom Gorman and Eric Lichtblau, "Anthrax Case Suspect Has Often Voiced Interest in Germ Warfare Arrest," *Los Angeles Times,* February 21, 1998, p. A21.

61 . . . *had spread the bacteria through ten local salad bars:* U.S. Senate Select Committee on Intelligence Hearing on the Threat of Biological Weapons, March 4, 1998.

61 . . . *Thomas Lavy:* Robert M. Burnham, Section Chief, Domestic Terrorism National Security Division, Federal Bureau of Investigation, statement before the U.S. House of Representatives Subcommittee on Oversight and Investigations, May 20, 1999.

CHAPTER 3

68 . . . *drawn from a 1999 presentation at a bioterrorism conference:* Thomas V. Inglesby, "Anthrax: A Possible Case History," *Emerging Infectious Diseases,* vol. 5, no. 4 (July–August 1999).

68 . . . *It would not be the first time that biological weapons were used:* Edward M. Eitzen, Jr., and Ernest T. Takafuji, "Historical Overview of Biological Warfare," *Textbook of Military Medicine,* Part One, "Medical Aspects of Chemical and Biological Warfare," Chapter 23; Office of the Surgeon General, 1997.

71 . . . *Modern societies are particularly susceptible:* Jessica Stern, *The Ultimate Terrorists,* p. 4.

71 . . . *where the objective for inflicting mass casualties can be obtained:* "Terrorism in the United States 1996," Federal

Bureau of Investigation, "Counterterrorism Threat
Assessment and Warning Unit, National Security
Division," p. 24.

72 . . . *a preponderance of urban Americans:* U.S. Census
Bureau, "Urban and Rural Population: 1900 to 1990,"
released 1995; available at *http://www.census.gov/population/
censusdata/urpop0090.txt.*

73 . . . MALL NEXT FOUR LEFTS: Joel Garreau, *Edge City:
Life on the New Frontier* (Doubleday, 1988).

73 . . . *why half of the Washington-area offices for the federal
government lie outside of the District of Columbia:* Francis C.
Turner, personal communication.

74 . . . *95,000 deaths and 125,000 people incapacitated:* Health
Aspects of Chemical and Biological Weapons (World
Health Organization, 1970).

74 . . . *These estimates are conservative:* Edward M. Eitzen,
Jr., "Biological Warfare: Historical Perspective";
available at *www.usamriid.army.mil/content/VioWarCourse/
HX-3/HX-3.html.*

76 . . . *Aum Shinrikyo:* Leonard A. Cole, *The Eleventh
Plague: The Politics of Chemical and Biological Warfare*
(Freeman, 1997), p. 152; compiled from the Japanese
press; staff statement, "Global Proliferation of Weapons
of Mass Destruction: A Case Study on the Aum
Shinrikyo," hearings before the Permanent Subcommittee
on Investigations of the Committee on Governmental
Affairs, U.S. Senate, 104th Cong., 1st sess., pt. 1,
October 31 and November 1, 1995, p. 66.

76 . . . *Impurities in the gas . . . kept the casualties smaller:*
Kyle B. Olson, testimony, hearings before the Permanent
Subcommittee on Investigations of the Committee on

Governmental Affairs, U.S. Senate, 104th Cong., 1st
sess., pt. 1, October 31 and November 1, 1995, p. 106.

CHAPTER 4

84 . . . *a staggering 76 million:* Paul S. Mead, Laurence
Slutsker et al., "Food-Related Illness and Death in the
United States," *Emerging Infectious Diseases,* vol. 5, no. 5
(September–October 1999).

CHAPTER 5

102 . . . *whiff of hysteria-fanning and budget opportunism:*
Daniel S. Greenberg, "The Bioterrorism Panic,"
Washington Post, March 16, 1999, p. A21.
103 . . . *The manufacturing technique is, in a sense, the real
weapon:* Kenneth Alibek and Stephen Handleman,
*Biohazard: The Chilling True Story of the Largest Covert
Biological Weapons Program in the World — Told from Inside by
the Man Who Ran It* (Random House, 1999), p. 97.
104 . . . *easy to produce:* Kenneth Alibek, prepared
testimony before the House Armed Services Committee,
Military Procurement Subcommittee, and Military
Research and Development Subcommittee, October 20,
1999.
105 . . . *Even groups with modest finances:* D. A. Henderson,
"Bioterrorism as a Public Health Threat," *Emerging
Infectious Diseases,* vol. 4, no. 3 (July–September
1998).
105 . . . *that tiny tube may contain enough biological material to
kill massive numbers of people:* Laurie Garrett, "Germ Terror:
Is U.S. Ready? Experts Worry Bioweapons Are a
Growing Threat," *Newsday,* April 6, 1998.

107 . . . *I have one hundred percent confidence that North Korea has it:* Kenneth Alibek, personal interview.

108 . . . *It was absolute bullshit:* D. A. Henderson, personal interview.

109 . . . *The only apparent security was one pimply-faced kid:* Dr. Peter Jahrling, personal interview.

110 . . . *the Iranian government had been working to recruit bioweaponeers:* Judith Miller and William J. Broad, "The Germ Warriors," *New York Times,* December 28, 1998, p. 1.

111 . . . *the scientists of "BIOEFFECT Ltd." are willing to sell their genetic engineering knowledge to anyone:* Kenneth Alibek, congressional testimony before the Joint Economic Committee, May 20, 1998.

112 . . . *nobody has stopped me:* William Patrick, personal interview.

113 . . . *would outstrip the technical capabilities of all but the most sophisticated terrorists:* Jonathan B. Tucker and Amy Sands, "An Unlikely Threat," *Bulletin of the Atomic Scientists,* July 1, 1999.

115 . . . *really is very, very simple:* David Pui, personal interview.

CHAPTER 6

126 . . . *Dr. John Bartlett walked into the emergency room:* All details drawn from John G. Bartlett, "Applying Lessons Learned from Anthrax Case History to Other Scenarios," *Emerging Infectious Diseases,* vol. 5, no. 4 (July–August 1999), and personal communications.

135 . . . *most state public health systems are unprepared to respond:* Ellen Gordon, testimony before the House

Subcommittee on National Security, Veterans Affairs and
International Relations, September 22, 1999.

135 . . . *few, if any, practicing clinicians have ever seen a
case of smallpox or anthrax or plague:* Tara O'Toole,
testimony before the House Subcommittee on National
Security, Veterans Affairs and International Relations,
September 22, 1999.

140 . . . *1972 smallpox outbreak in Yugoslavia:* D. A.
Henderson, "Bioterrorism as a Public Health Threat,"
Emerging Infectious Diseases, vol. 4, no. 3 (July–September
1998).

142 . . . *won't deliver the goods without wasting an inordinate
amount of money:* Richard Preston, "Demon in the
Freezer," *New Yorker,* July 12, 1999, pp. 60–61. The
situation with an anthrax vaccine is also troubled.
The military continues in its massive program to
vaccinate all 2.4 million of its troops against anthrax,
which requires a series of six shots per recipient. The
only licensed U.S. manufacturer of the vaccine, Bioport,
enjoys an exclusive $49-million contract with the Defense
Department to manufacture and store the vaccine that
was granted to a company Bioport bought, Michigan
Biologics Products Institute (MBPI). That company was
once part of the Michigan Department of Public Health.
MBPI's manufacturing practices drew criticism from the
Food and Drug Administration, which suspended the
company's vaccine approval in March 1997 until it improved
its processes to ensure the quality and potency of the
vaccine. In February 1998 the FDA ordered eleven lots
of the company's anthrax vaccine withdrawn because of
further potential problems with potency and sterility of

the vaccine. Bioport was created in part by Admiral
William J. Crowe, former chairman of the Joint Chiefs of
Staff, and business partners for the purpose of taking over
MBPI. MBPI halted production of anthrax vaccine in
January 1998 to begin a comprehensive renovation of the
anthrax production facilities. "No anthrax vaccine has
been produced since Bioport became owner of the facility."
FDA's director of the Center for Biologics Evaluation and
Research, Kathryn C. Zoon, testified before Congress that
by the end of 1998, Bioport's facilities had improved. FDA
Website (*www.fda.gov*), "FDA Warns Michigan Biologic
Products Institute of Intention to Revoke Licenses";
Howard L. Rosenberg, "Anthrax Cloud's Silver Lining:
Bioport Corp. Lands Exclusive License to Produce
Vaccine," *20/20*, March 12, 1999; Kathryn C. Zoon,
testimony before the Subcommittee on National Security,
Veterans Affairs and International Relations Committee on
Government Reform, U.S. House of Representatives, April
29, 1999; available at *www.fda.gov*.

142 . . . *much of the supply might not be usable:* Laurie
Garrett, "Smallpox Vaccine Tainted — U.S. Protection
Threatened," *Newsday*, April 23, 1999, p. A8; Dr. James W.
LeDuc and John Becker, letter to *Emerging Infectious
Diseases*, vol. 5, no. 4 (July–August 1999).

CHAPTER 7

149 . . . *A bioterrorist event is different from all other forms of
terrorism:* Margaret A. Hamburg, testimony before the
Senate Committee on Health, Education, Labor and
Pensions, Public Health Subcommittee, March 25, 1999.

151 . . . *Alive, maybe:* Tom Clancy, *Executive Orders*
(Putnam, 1996).

151 . . . *These are minimum estimates:* Arnold F. Kaufmann,
Martin I. Meltzer, and George P. Schmid, "The Economic
Impact of a Bioterrorist Attack: Are Prevention and
Post-Attack Intervention Programs Justifiable?"
Emerging Infectious Diseases, vol. 3, no. 2 (April–June
1997).

152 . . . *President's Commission on Critical Infrastructure
Protection:* "Critical Foundations: Thinking Differently,"
The President's Commission on Critical Infrastructure
Protection; available at *http://www.info-sec.com/pccip/web/
report_index.html.*

152 . . . *A prominent legal scholar who has examined the issue:*
Much of the material in this chapter is drawn from a yet-
unpublished paper by Terry P. O'Brien, "Legal Response
to a Bioterrorist Event," and from private
communications with O'Brien.

157 . . . *consequence management is the crisis:* Terry P.
O'Brien, quoting Chris Seiple, "Consequence
Management: Domestic Response to Weapons of Mass
Destruction," 119 *Parameters* 122 (Autumn 1997).

157 . . . *a vague document that provides no blueprint:* Richard A.
Falkenrath, Robert D. Newman, and Bradley A. Thayer,
*America's Achilles' Heel: Nuclear, Biological and Chemical
Terrorism* (MIT Press, 1998).

159 . . . *A 1984 disaster-relief law:* Robert T. Stafford,
Disaster Relief Act of 1984, 42 U.S. Code sec. 5121 et
seq.

160 . . . *rules of engagement:* Thomas R. Lujan, "Legal

Aspects of Domestic Employment of the Army,"
Parameters, August 1997.

164 . . . *federal, state, and even local governments don't get it:*
After the B'nai B'rith incident, the number of hoaxes rose
sharply, leveled off briefly, and now appear to have begun
rising again. In 1999 (through late October) the FBI
investigated 225 new cases of possible chemical and
biological terrorism nationwide—an increase over the 181
new investigations that the Bureau conducted all of the
previous year, according to a report in the *Washington
Post*. The FBI's special agent in charge of the FBI's
national capital domestic response team, Jim Rice, told a
group of Washington-area police and fire officials that
downtown Washington alone receives three to six
suspicious packages a day that must be handled carefully
by experts trained to detect explosives as well as chemical
and biological weapons.

Although the vast majority of incidents into the future
will be hoaxes, even false reports of attack can be both
useful and damaging, HHS official Hamburg says.
Hoaxes, she says, "offer an opportunity to examine
our coordination and response. Thinking through the
different types of hoaxes helps us develop protocols and
strategies that lead to recognition of a true event." On the
other hand, after responding to or hearing about many
inconsequential hoax situations, it's easier to take
bioterrorism for granted.

You might accept the notion put forward by some
critics of the movement to fund civil biodefense that
hoaxes are likely to be the worst we will ever see of
bioterrorism because of the difficulty of mounting an

attack, or that the hoaxes themselves lead us to overestimate the threat and to spend too much on biodefense. As I show in Chapter 8, the biggest problem is not that money is being spent — but that much of it is being spent on the wrong things.

CHAPTER 8

167 . . . *The once improbable has become the inevitable:* Congressman Christopher Shays, prepared statement, House Government Reform Committee on National Security, Veterans Affairs and International Relations, September 22, 1999.

168 . . . *what followed was* really *frightening:* In fact, the program contained inaccuracies that D. A. Henderson, in a letter to Kyle Olson, the expert consultant for the program, called "regrettable." Most important, the program treated the life cycle of the epidemic as a mere eight days, when evidence from the 1979 Sverdlovsk epidemic in the Soviet Union showed that the release of anthrax spores generated cases for as many as forty-seven days after the initial exposure. The point would be crucial in preparing a response to an outbreak, since it means that rapid availability of antibiotics to the exposed population could actually save lives. The fictional eight-day scenarios used by *Nightline* would render any antibiotic prophylaxis use moot; the population would be dead before you could even get them the pills. "Given the current plethora of inaccurate and misleading information, it is unfortunate that an opportunity such as was presented could have been so irresponsibly handled." D. A. Henderson, private correspondence.

171 . . . *the U.S. Government is not effectively organized to combat proliferation:* "Combating Proliferation of Weapons of Mass Destruction," report of the Commission to Assess the Organization of the Federal Government to Combat the Proliferation of Weapons of Mass Destruction Pursuant to Public Law 293, 104th Cong., 1999.

172 . . . *a General Accounting Office report:* GAO, "Combating Terrorism: Issues to Be Resolved to Improve Counter-Terrorism Operations," *NSIAD-99-135,* May 13, 1999.

173 . . . *no "silver bullet":* Robert F. Knouss, prepared statement before the House Government Reform and Oversight Committee, National Security, International Affairs and Criminal Justice Subcommittee, September 22, 1999.

174 . . . *What are you going to do with all that money?* Margaret Hamburg, private interview.

174 . . . *a real spree:* Paul Jacobs, "Attack of the Killer Microbes," *Los Angeles Times,* August 19, 1999, p. A1.

176 . . . *the bio-terrorism arena:* Patty Reinert, "Tech Center Funding Vote on Tap Today: Research Would Target Bioweapons Terrorism," *Houston Chronicle,* October 13, 1999, p. 25.

CHAPTER 9

203 . . . *there are more things to admire in men than to despise:* Albert Camus, *The Plague* (Knopf, 1991), pp. 286–87.

INDEX

ABOUT THE AUTHORS

MICHAEL T. OSTERHOLM, PH.D., MPH, the former Minnesota State Epidemiologist and former Chair and CEO, ican, INC., has been an internationally recognized leader in the area of infectious diseases for the past two decades. He is the recipient of numerous honors and awards from the CDC, NIH, FDA and others, and served as a personal advisor on bioterrorism to the late King Hussein of Jordan. He has led numerous successful investigations into infectious disease outbreaks of global importance. A frequent lecturer around the world, he is now director, Center for Infectious Disease Research and Policy, and professor, School of Public Health, University of Minnesota.

JOHN SCHWARTZ is a reporter at *The New York Times*; he writes about technology and business and their impact on society.